WYRE FOREST REFRESHED

WYRE FOREST REFRESHED

The Autobiographical Notes of a Biologist

Norman E. Hickin

Ph.D, B.Sc, F.R.E.S.

Kenneth Tomkinson Limited

1981

ISBN O 907083 02 1

Kenneth Tomkinson Limited
High Habberley House
Kidderminster, Worcestershire, England

To Emma

my wife

whom I first met on the bank of

Dowles in Wyre Forest

Contents

Preface

There are two themes to this book. In the first place there is the autobiographical theme. I have always been intrigued by the difference between people in regard to their natural inclinations. What makes us want to do the things we do? What factors go towards influencing any choice we have in selecting the profession or kind of work we perform in the social structure? What I have written in this book explains *none* of these enigmas, but what I have endeavoured to do is to record my own burning desire, from my earliest recollections, to steer a certain course and to become knowledgeable on a subject which for a long time appeared to have little practical significance in my life.

I wanted to be a biologist—in general, a naturalist. More specifically I wanted to be an entomologist. I wished very earnestly to make my living by a knowledge of insects and their ways. The first part of the book, then, is largely concerned not only with my recollections of the various natural phenomena that perhaps provided a stimulus but how I went about obtaining the essential qualifications, academic and otherwise.

I have recorded also my reactions to many topics and incidents so that the intensity of my feeling for my profession may be gauged. There is, of course, a wide variation amongst people in the intensity of any feeling in this matter, and, of course, in the amount of time which any man is prepared to give to his professional or vocational pursuits. Fortune has, indeed, been very kind to me in several ways. From my earliest days up to the present, men and women of outstanding attributes have been kind enough to influence and encourage me to do what I wished to do. Of equal importance, even if less tangible, has been the strength that has been given to me by a piece of countryside. This at first sounds ludicrous, but, of course, this is by no means a unique suggestion and it is the second theme of this book.

The quietness, the sense of being apart from all other humans, except one's loved ones, and yet seeing natural things all interlaced in a pattern, a pattern which has been replicating for hundreds of years with but little change, all contribute to strengthen the resolution. Seeing unusual and often exciting animals and plants each year repeating their characteristic story has always given me refreshment. Of latter years I have called on this to a much greater extent so that I am now a long-range commuter, driving nearly 150 miles each Friday evening and again on Sunday evening so that I may be forest refreshed.

Acknowledgements

In a sense this book is wholly an acknowledgement to all those who helped me to become a naturalist and a professional biologist. In this one person stands out above all others in encouraging me to go out into the fields and woods and observe living things, to go to the libraries and read about them, to write about them and to draw them. My mother, of course. There was no special reason why, from her own background, she should want to do this. It was just her intuitive rightness.

I mention a number of men and women by name who have helped me but there were many others who for various reasons I do not mention. I hope any of the latter who read this will forgive me.

Of the seventy-six illustrations in the book, I am responsible for seventy-two. I was much taken by the drawing style of Mr. Mortiboys of my old school, Aston, and he drew figure 2. Figures 71 to 74 inclusive are taken from the old books in order to document the unique tree of Wyre Forest, the Whitty Pear.

Miss Daphne Linscott read and corrected the typescript and made many useful suggestions. I am very grateful to her.

Lastly, I acknowledge the help of my wife Emma in making it possible for me to spend such a high proportion of my time writing and drawing. During the last twenty-five years I have made well over a thousand drawings of insects which have been published to illustrate my scientific papers and these have all been done in my wife's company during evenings and weekends.

I

A Glimpse of Nature

THE coping stones of the wall were blue, shiny and triangular. Their broad bases lay on the grey dusty bricks that were once a light coral pink suffused with buff patches. The top of the copings came to a sharp point. But under the copings was a dimly lit corner which ran the whole length of the wall, which I should say was just about a yard high. Where the mortar had flaked away here and there were buff and gritty patches and it was in these small crumbling areas, like broken blisters, that the harvestmen *Phalangium opilio* lay. They were blister-like themselves with their eight very long jointed legs spread out like a spider's web, or grouped together into fissures in the wall. The mottling on the back and the irregular placing of the legs made these creatures hard to find, so well did they fit into their background. I know the species now to be members of the *Phalangiidae* family, allied to, but not true, spiders. When I collected them I did not know their species. Not to be wondered at, perhaps, as I was only three years old at the time, but I placed them amongst the spiders. My mother records the answer I gave to the person enquiring as to what I was doing. 'I finding 'piders, I no kill 'em,' was my reply. Later I gave them the name of daddy-longlegs, a name more correctly to be given to the crane fly.

My mother, Alice Priddey, was born in 1887. She had an elder brother, two younger brothers and two younger sisters. Her father, James Priddey, was of Gloucestershire stock, his father coming from Standish Park near the south bank of the river Severn. My great-grandfather, also named James Priddey, came to Birmingham as a young man and as a cog in the industrial revolution. My grandfather was a handsome man and of substantial proportions. He spent all his working life as an engineer with the Birmingham manufacturer of minting machinery, Taylor and Challen, as did his three sons, all now retired. He worked a lathe so large and requiring such physical strength that on his retirement, due to the onset of blindness, no one could be found to follow him and the lathe was dismantled.

Although he wore overalls, he made a fine collection of oriental china and was considered an expert on the subject. He was a capable musician, playing the violin and having a wide knowledge of violins and their makers. Two of his sons made violins as a leisure-time occupation. Sunday mornings were spent playing chamber music, string quartets and piano quintets. My mother had spent a great deal of her youth at the piano, as my grandfather had thoughts of his daughter becoming a concert pianist, and when she wished to marry at the age of twenty he was very upset. My mother was a very capable pianist and, indeed, still is to this day, and my brother and I must be very grateful to her for her earning power during the years she played the piano at cinemas in the 'silent' days. She was extremely good at this. Not only was she able to play a great amount of music by heart but she had a keen sense of fun. She would watch the picture on the screen, play the piano with her right hand, whilst with her left she would snatch a piece of music she was playing, put it on a pile at her side and replace it with a piece from another pile. I have never seen such musical dexterity since, but, of course, such music juggling is no longer required.

My mother has shown great dexterity in other directions, more especially with the needle and the sewing machine. Until I was eighteen years old she made every garment that my brother and I had ever worn, except for underwear and socks. Shirts with attached collars, grey flannels, tweed jackets or tweed overcoats she would 'run up' with great speed and with professional finish. I was never conscious of wearing 'home-made' clothes; indeed, I was always rather proud in a general sort of way of my clothes, because I knew the cloth was of the highest quality.

She made me about a dozen tents in all, for our camping expeditions, but one of her specialities was the production of composite blankets or bed-covers made in her own type of patchwork. One cover, now in the possession of one of my daughters, is made up from family historical material. It consists of pieces taken from my old rugger socks, my grammer-school cap, the flannel shirt I was wearing when I met Emma my wife, and the first dress I bought her after we married, etc., etc. She made a dozen or so of this type of thing. Another speciality of hers, made on the sewing machine, were bags of various sizes made of a special sort of plastic-covered cloth. She made eight of one design for my family safari that were virtually dustproof, weighed less than 2 lb. each and carried 22 lb. of luggage, so that two carried 44 lb.—the tourist-class weight limit in aircraft. They have since been all over the world and have stood up superbly to the hard wear of foreign travel.

My father was a very steadfast man. He was the most just, honourable and reasonable gentleman that I ever met. His name was Ernest Allen Hickin and he was born in 1885 in Birmingham. His father was a master brass-founder whose business still prospers in Birmingham under the direction of a grandson. When my father was sixteen my grandfather died, and due to a family upset, my grandmother and her three sons made preparations to travel to Australia, leaving the children of two previous marriages in charge of the family business. However, she married an insurance agent and settled in Rubery just outside Birmingham in Worcestershire, whilst two sons went off to Australia without her.

My father was apprenticed to cabinet-making, a craft which he followed almost all his life. The first impression of him is in 1916 when I was six. He enlisted in the Devonshire Regiment and I remember his blue-grey eyes, long pointed wax moustache, like Kitchener, a very straight back and sandy hair. He married my mother when she was twenty and he was twenty-two, and he gave her his complete and absolute devotion until he died at the age of seventy-seven. Throughout their lives together they were intensely happy in each other's company and unashamedly displayed affection for each other in front of my brother and I. I remember, too, that, he smelt of oak and glue and had one of the most melodious tenor voices I have ever heard.

When I hear the angels I can hear my father singing duets with my mother on Sunday evenings or at parties and I can see him with his head thrown back and his soul shining through his eyes. 'Oh! that we two were maying', and our favourite solo, 'Take a pair of sparkling eyes'. He sang as though the songs were specially written for him alone. He enjoyed playing the 'cello, gardening and listening to the football results on the radio on Saturday teatimes, although if Aston Villa had lost he might switch off before the end.

From the time he was a young man until his death he was an unwavering socialist imbued with the spirit of William Morris and until he died, able to quote from the articles of Robert Blatchford.

When my father and mother married they rented a very small terraced house, 73 Alma Street, Aston Manor, then in the county of

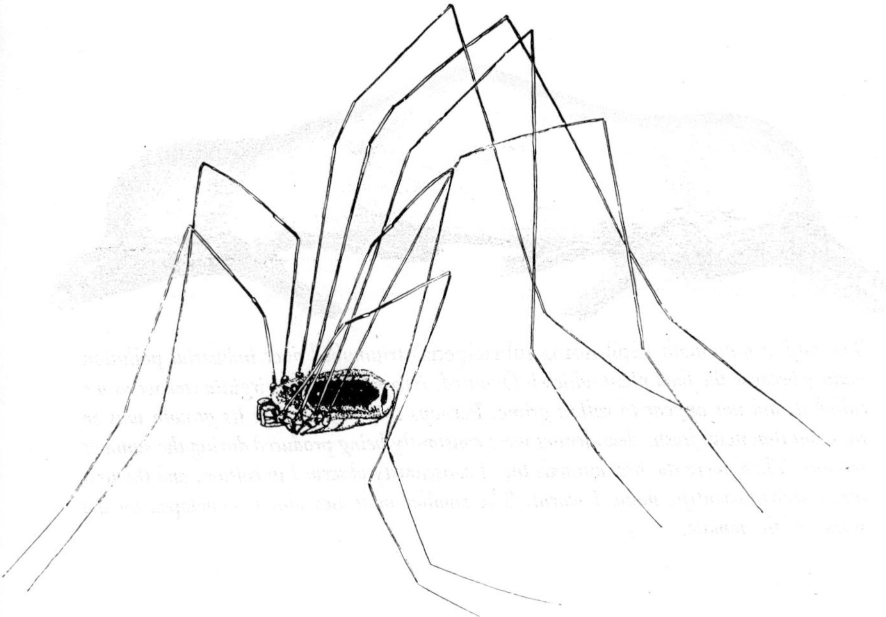

The harvestspiders are only distantly related to the true spiders but both are placed in the class Arachnida. *The true spiders have the body divided into two distinct parts, the anterior of which is known as the cephalothorax and which bears the organs associated with the head and the eight walking legs. The posterior part is the abdomen. In the harvestspiders, however, of which there are twenty-one British species, the cephalothorax and abdomen are joined together without the distinct 'waist' shown in the true spiders.* Phalangium opilio *was the first creature, a harvest-spider, which excited my interest.*

I was born at 73 Alma Street, Aston in Birmingham, in 1910. There were no trees, there was no grass, not even a garden of any sort in which to grow some industrial grime-resistant weeds. Perhaps the force which now compels me to seek happiness in the forest's secluded places has its origin in Alma Street. (J. Mortiboys.)

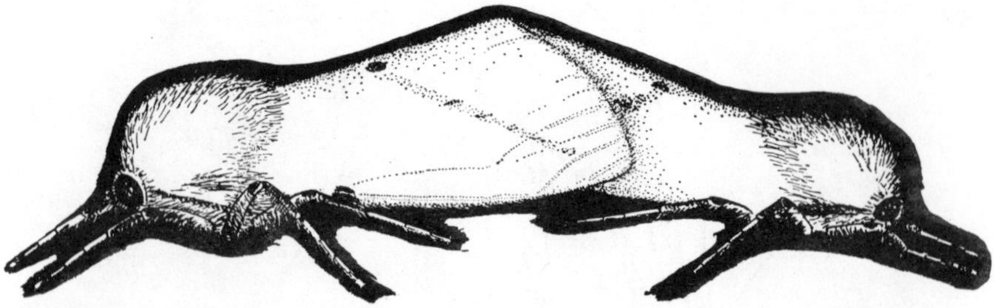

The buff ermine moth (Spilosoma lubricipeda) *triumphed over industrial pollution mainly because the food plant which it favoured,* Ampelopsis, *or virginia creeper as we called it, did not appear to collect grime. Perhaps this was because its growth was so rampant that new, fresh, clean leaves were constantly being produced during the summer months. These were the first animals that I consciously observed in coition, and the first insect whose scientific name I learnt. The smaller male lies almost enveloped by the wings of the female.*

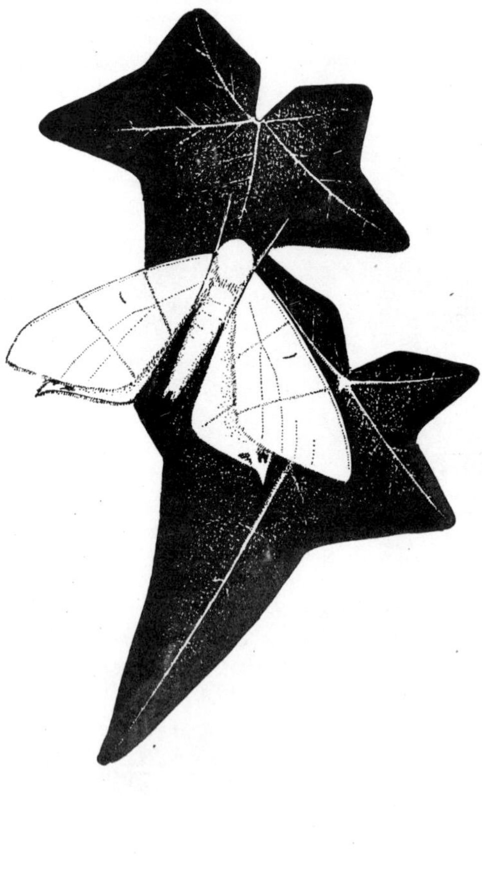

Nearly all the larvae of the moth family Geometridae *bear a close resemblance to a short piece of stick often complete with little knobs and excrescences. They are usually called stick caterpillars or 'loopers' on account of their peculiar method of progression. The adult moths are thin-bodied with large rounded delicate wings. From this it must not be assumed that they are weak fliers. The light silvery yellow colouration and almost white body of the swallowtail moth when at rest gives it an ephemeral appearance, but it has a strong fluttering flight. Today when my eye chances on it my memory takes me back to Birmingham as this was the first geometer that I encountered.*

Staffordshire but shortly afterwards taken into the county of Warwickshire by the city of Birmingham. The district was mean and disagreeable and shortly after I was born, on 17th April 1910, my parents moved to Wilson Road, Birchfields, about half a mile to the north of Six Ways, Aston. Experts in the niceties of Birmingham dialects can recognize mine with great accuracy as 'Lozells'. The house in Wilson Road had practically no garden and a saw-mill abutted to the rear, but the few square feet of the front garden was filled with an immigrant weed whose identity I have never been able to establish. It was a balsam, rather like the North American orange balsam *Impatiens biflora*, but with larger maroon-coloured flowers.

My brother Kenneth was born on 4th January 1912, and soon afterwards we moved to Putney Road about half a mile away, and it was here that I commenced to notice the natural history of Handsworth. I can recall the first moment when the striking beauty of insects first struck me. A whitish speckled daub on the window suddenly revealed itself to me as a pair of buff ermine moths *in delicto flagranti*. (This was the first time that I had ever seen two animals pairing.) The wonderment produced in me by the revelation of their immaculate scaling and the delicacy of their hairiness has lasted to this day. The perfection of a newly emerged insect, the minutiæ of pattern on an insect's egg or the critical balance of a parasite and its host have always had a similar effect on me.

As soon as I started at the primary school my parents made a butterfly net for me (and one was made for my brother too). Father carved a piece of oak in the shape of a handle with the top extending outwards and in the wing-like extensions long holes were bored into which the ends of the piece of cane fitted. The natural resistance to the flexibility kept the cane from flying out. My mother, of course, made the bag and it was a very good net. The nets that could be purchased from Spicers, the naturalists sundriesmen and taxidermists in Suffolk Street, had a Y-shaped handle made of brass tubing soldered together and the cane was jointed in three places with brass ferrules and slides. One was once bought for me and I lost it the first day out. I cannot recall losing anything since that day, but the fact that I once *even* lost my butterfly net is given as an example by my family of my habit of losing things.

Just at what age I first found the swamp I cannot recollect, nor can remember my first visit to it, but I was certainly very young and wa attending the local council primary school at nearby Westminster Road. Invariably Kenny would be my companion and we would set off each with a cotton khaki haversack containing our corned-beef sandwiches and currant cake and some empty tobacco tins. We had a choice of routes: we could make our way to the Birchfield Road down which clanked the noisy No. 6 trams to Perry Barr, thence along the Walsall Road and down Rocky Lane; alternatively, we could pass through the Handsworth Wood area and down Hamstead Road to Great Barr station, where just short of the colliery we turned along an ash track back along the railway line until we came to our Eldorado.

The swamp covered about fifteen acres. The railway ran along one side, a heavy creosoted wooden fence divided it from the ashy track, then along one corner ran the river. It was as black as charcoal, and stank, and when we fell in it, as we did quite often, the smell lingered for a long time. But in this waste corner there was never interference with the plant life, with the beds of water mint attracting a large number of butterflies. It was there I first saw the small tortoise-shell, the painted lady and the clouded yellow amongst the purple balls of the water-mint flowers.

It was from the swamp that I first collected aquatic animals with the intention of trying to keep them alive at home. Among the first of my prizes was the large and wonderful *Dytiscus marginalis* water beetle. I kept still for an agonizing time around a little pool with the net poised, waiting for this aquatic beetle to exhaust its supply of oxygen. During this time my intended victim went swimming with its oar-like legs along the bottom of the pool, which was too murky and brown for me to follow its course. The beetle eventually floated to the surface bottom up, where, with a quick and darting stroke of the net, I secured it. Its four-inch-long larva was also found with its sickle-like jaws ready to stick into everything that moved, although tadpoles were its usual diet when I kept it in captivity.

Then, over by a row of poplar trees quite close to the river, was a patch of ladies bedstraw which scrambled around the rushes. It was a most exciting day when I saw the huge larvae of the elephant hawk moth in great profusion. They were mottled and creased like elephants' trunks and when disturbed they quickly pulled in the head and the two first small segments of the body, to give a realistic impression of a snake with head and prominent eyes. Some of my time was always spent poking in the water, and dragonfly larvae, including the large *Aeschna*, were found. It was here that I first found caddis worms, the larvae of caddis flies that wrapped a mantle of vegetable nondescripta around their white and naked bodies.

On the far side of the swamp was a row of small cottages with a pump on a lawn of short geese-grazed grass. We would sometimes see two or three coal-black miners walking back to their cottages from the nearby Hamstead Colliery with its great cone of 'dirt' dominating the skyline. But adults never interfered with our ploys at the swamp, though sometimes we were chased out by belligerent youths, and twice there were fights which somehow made our visits to the swamp rather like pioneering into a strange country where we must always be ready either for flight before superior weight of unfriendly natives or to defend ourselves.

In the great sea of working-class suburbia that was Handsworth, to the north of the centre of Birmingham, the flora was distinctly limited. An extremely high proportion of the little front gardens had waist-high brick walls encircling them and behind the brick wall a very dusty but nevertheless vigorous privet hedge. In the evenings of high summer the hedges were clipped with shears almost as a ritual, and although the privet was so dusty and dirty it supported a very substantial population of a moderately large fat-bodied geometrid moth, the scalloped hazel. The adult moths were seldom seen, but an hour's careful search in autumn would usually yield about a dozen stick-like larvae, and for several years I reared a few moths out in a very beautifully made rearing cage, my first piece of specially purchased entomological apparatus supplied by Spicers of Suffolk Street (for only 6*s.* 6*d.*). At the end of the road about 200 yards away grew some poplar trees and each year a few poplar hawk moths were found round about. I was ten years old when I collected my first pair, but when they were pinned out on setting boards the cat tore them up, evidently thinking that the large furry bodies had a mouse-like quality.

Ivy was another plant that would flourish under the dusty conditions of the Birmingham suburbs and this supported numbers of the beautifully delicate and light silvery yellow swallowtail moth. I had difficulty in finding the caterpillars, but the adult moths flew at dusk on hot July nights along the rows of front gardens with incredible swiftness. My mother caught my first specimen for me after I had been sent to bed as being past bedtime.

* * * * *

To many thousands of children living on the north side of Birmingham, Sutton Park, an ancient gift of Bishop Vesey to the royal borough of Sutton Coldfield, opened up a new avenue of happy adventure. The first fifty times or so I made the five-mile journey it was by foot, but thereafter a series of old bicycles carried me. We walked through the brick and slate of Perry Barr and then by the tired and dusty fields of Kingstanding until, quite suddenly, the outlook changed and we were no longer in the grip of the town; we were in the country. The sandy, undulating lane between the high hedges to the park at Banner's Gate often gave us something of interest and it was there on a sandy bank under a holly tree that I first realized that the bee world held something other than the humble bee and honey bee. Each year a number of small rusty-coloured Andrenid bees made themselves busy tunnelling into the sand at this place, the little piles of dark freshly excavated material showing up distinctly against the light surface sand.

At one point along the Banner's Gate Lane a group of large lead-grey barked beeches stood by a wide bare verge. Here squatted the stone breaker by a pile of stone that would have lasted him his life in this world and the next world too. I remember him as being singularly uncommunicative.

After breasting the top of the rising lane Kenny and I would almost run down to the main Chester Road which skirts the park at this point, then across to the keeper's lodge where we paid our pennies for admission. Tenants of the royal borough paid nothing for admission. This was always a point of contention with Birmingham citizens that Sutton residents had all the advantages of the municipal services which the great city had to offer and gave nothing in return, paid lower rates and then charged us to go into their park. To two small boys, however, to have the pennies was definitely something! Usually we went on, keeping Longmoor coppice on our left and picking out steps across the bog at the head of Longmoor pool. It was here that I first met the curious entomophagous plants sundew, *Drosera rotundifolia*, and butterwort, *Pinguicula vulgaris*, but I doubt very much whether they still exist there.

The holly blue was the first of the 'blue' family I caught. It was very common in Sutton Park; there must be more holly to the acre there than anywhere else in the country. The holly blue is an early flier, usually appearing during the first few warm and sunny days of April, and I used to catch it flitting round the holly hedge of the lodge at

'There are fairies at the bottom of our garden,' I used to recite when I was eight years old. But I knew that there were 'angels' at the bottom of the garden next door. This was the name we gave to the snow white plume moths (Alucita pentadactyla), *that were easily disturbed from amongst the thick enveloping foliage of the large white trumpeted convolvulus. In the early dusk of high summer I would watch them make their feeble flight of three or four feet at most before they settled back on to the broad spear leaves. On alighting they immediately took up a rest posture with the antennae tucked under wings which are spread out transversely in such a way that the hind pair are invisible and the long spurred white legs take a characteristic lace-like pattern.*

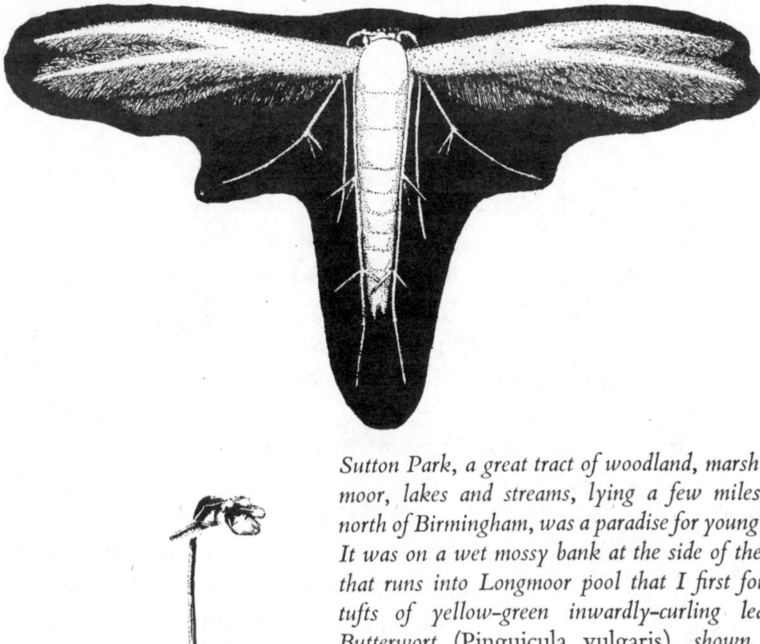

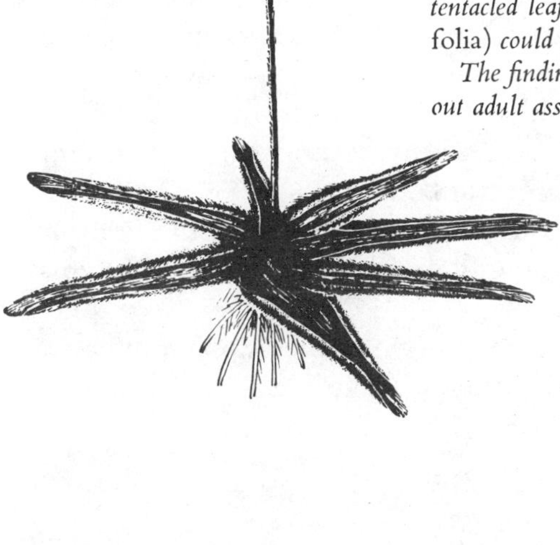

Sutton Park, a great tract of woodland, marsh, heath, moor, lakes and streams, lying a few miles to the north of Birmingham, was a paradise for young people. It was on a wet mossy bank at the side of the stream that runs into Longmoor pool that I first found the tufts of yellow-green inwardly-curling leaves of Butterwort (Pinguicula vulgaris), *shown in the illustration with its violet-like flower. The red-tentacled leaf rosettes of Sundew* (Drosera rotundifolia) *could be found there also.*

The finding of these two insect-eating plants, without adult assistance when I was ten years old, thrilled me to such an extent that whenever I see these plants in the boggy lonely places in our islands I become almost transfixed and I move away with reluctance to catch up with my family. Sundew used to occur on a marshy hillside near the island in Wyre Forest on the Salop side of Dowles Brook, but I have not seen it there for many years, the marsh having dried up. Butterwort is also recorded from the forest but I have never observed it there.

Banner's gate at the very entrance to the park. I had caught about a dozen of this species before I left the primary school at the age of eleven.

I can remember very clearly where I caught the first specimen of all the species of butterfly that came to my net and the exact circumstances of the capture. The meadow brown, small heath and the pearl-bordered fritillary were captured in the Rose and Crown Gardens at the Lickey Hills. The speckled wood butterfly I caught in the woods between Borth and Aberystwyth. Looking over a stone wall near Conway in North Wales when I was six years old I saw three peacock butterflies flitting round some thistle heads. The little blue, the chalkhill blue and the adonis blue I found first on Cranham Bank when I was sixteen.

In the summer of 1916 I went to stay with an aunt in the village of Gyffin near Conway in North Wales, and on a rocky hillside I saw my first grayling butterfly. I was six years old, so that I was lacking both in speed and finesse. It is not to be wondered at, therefore, that they nearly eluded me. One would fly up at my feet and go hurtling across the rough ground and I would make my way to the spot I had marked down with extreme caution, and search the ground for the butterfly. Then suddenly it would fly up and go zigzagging away over the rocks and again I would follow up. This happened many times before I could make out the insect lying flat against the lichened rock. I was then able to collect two or three before making my way back to Gyffin, happy not only because I had collected a new species of butterfly but also because I had learnt one of its habits, that of tipping over to lie with its speckled and blotched wings flush against a matching background so that it became virtually invisible.

Another very common butterfly at the Great Barr swamp was the diminutive small copper. The larval food plants, sorrel and species of docks, were present in abundance and since those days I have never seen this butterfly in such numbers. It is an insect which is well known to vary in abundance from year to year and as there are three or, in some years, four life-cycles in the year, it can be seen from May to October. At the swamp it was to be seen on its favourite flower, the fleabane, although during the autumn it certainly liked to sit upon a michaelmas daisy in gardens.

I also had an early acquaintance with two of the skipper butterflies, the little grizzled and the brown dingy. These May butterflies were

The grizzled skipper butterfly, Hesperia malvae, *loves the hot sheltered banks where the wild strawberry,* Fragaria vesca, *grows and at the end of April when it commences to flower so the butterfly appears. The flight is quick and skippy but it does not travel far for it flies in a circle, usually returning near to the place from which it was disturbed, but when a pair are courting only the quickest eyes can follow their cavorting.*

Industrial development at the northern edge of the city left large areas of waste land by the side of the river Tame. It became useless to man and beast and the couch grass invaded it, which in time produced a huge acreage of monoculture. The ghost moths, whose larvae feed on its roots, followed, and in the dusk of summer by the stinking and black river the silver males flitted amongst the long spiky blades of couch seeking the virgin canary yellow and rufous striped females sitting on the ground after their emergence.

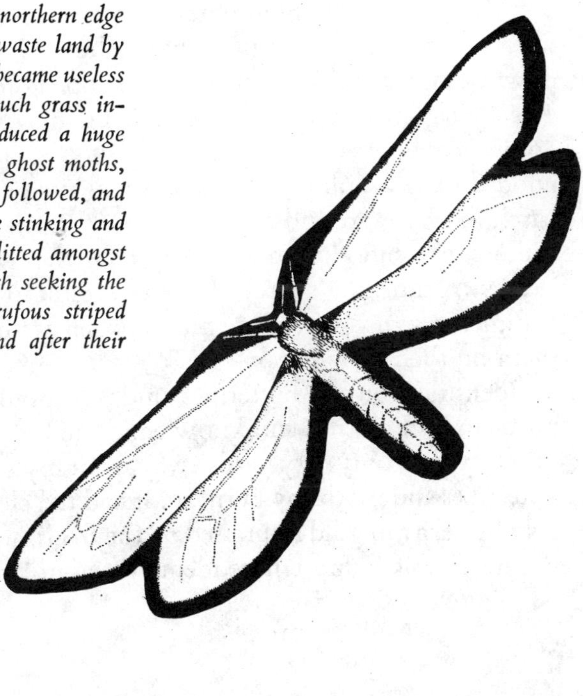

common in the flowery rides in Cofton Woods near the Lickey Hills. The grizzled skippers were very difficult to catch, but I had a quick eye, and the little collections of butterflies that I made for my council school at Westminster Road, Handsworth, contained a bright little pair as well as a clouded yellow, a migrant which was extraordinarily abundant in 1919. This black-edged brilliant yellow butterfly could be seen everywhere in the later summer of that year.

The common blue butterfly I first saw on the railway bank between Great Barr railway station and the swamp where I was so happy as a small boy. I was eight years old and had to get through the wire of the railway fence to net a few specimens on the bank near the railway lines. I was frightened and glad to get back through the wires again on to the dirt track. Also in the swamp the first small tortoise-shell butterflies were caught, as they probed the mauve pom-pom-like flowers of the water mint. The green-veined white I also first caught at the swamp. The cabbage whites, the small and the large, were, of course, the first butterflies I remember seeing. They were very common, very much more numerous than they are today, I believe. The red admiral was the next butterfly that I noticed, congregating around a willow tree in a front garden, attracted by the honeydew secreted by aphids. The sticky layer on the leaves was covered by grey dust, giving the tree a strange appearance. Any brightly coloured butterfly or moth was known as a bob'owler by my contemporaries, whether a red admiral butterfly, a tiger moth or a yellow underwing moth. It had to be pursued until downed under half a dozen jackets, stripped off quickly for the purpose. The appearance of the specimen can be imagined after such a turmoil. Sometimes white butterflies were pursued by the boys of the neighbourhood during a holiday. They were awaited on the 'old peck', a bare patch of waste ground at the corner of Putney and Crompton Roads, and as the white fluttered its way over the 'peck' it was assailed with jackets. A morning's catching would result in a jam jar half full of the mangled remains of two or three dozen with hardly a scale between them on their now greasy-blue wings.

Identification of butterflies and moths did not appear to be a problem. A tall, thin and knowledgeable boy who lived at the corner of the avenue only fifty yards away from my own house had a book on butterflies and moths by Furneau. It had red cloth covers with a painted lady butterfly in gold impressed on the front, and it was often consulted. If a moth was found whose identity was not known, our book-owner would run and fetch the coveted book and we would all crowd round it on the pavement, the owner just keeping a precarious hold. Only a few moth species were likely to be found in our road, so that I think we were usually right. The yellow underwing and the garden tiger were very common, although I have not seen one of the latter for many years. Forty years ago they were present in extraordinary abundance, the larvae, which we called 'woolly bears', being on every dock plant. In the area of Perry Barr and Great Barr the large reticulation of the brown and white fore-wing pattern varied greatly and many extraordinary varieties originated from this area.

For the Christmas of 1922, when I was twelve years old, my mother and father gave me Kirby's *British Butterflies and Moths*. It was packed with watery-coloured plates and cost them 10*s*. 6*d*., a quarter of my father's weekly wage at that time.

Insects, other than butterflies and moths, were difficult to identify without books to hand, and reliance often had to be placed on the spoken word of adults. There was, however, a branch of the Birmingham library in Perry Barr, and, although a very long walk for a small boy, I borrowed in turn every book on the natural history shelves.

A botanical garden was situated at Edgbaston on the more expensive side of the city. There was also a collection of animals, including a brown bear and obscene-looking pink and white axolotls. But the sight that thrilled me were the grey squirrels in the trees. It was explained to me that they had been introduced into Britain from North America, had driven the indigenous red squirrels away and had now become a menace to horticulture and our wild life; this in 1920.

This, of course, was before the 'open access' system. I would go to the polished counter, so high I could not see the rows of books behind, and when my turn came at last I would ask for a book on Natural History or birds or plants or Africa. The assistant would walk off to the appropriate shelves and bring something back to show me to see if I had read it before. This was time-consuming by modern standards, but one bookcase had a sheet of glass in front so that if you were adult or otherwise tall enough you could put your finger on the glass opposite the book you wanted. I often borrowed Miall's *Natural History of Aquatic Insects* until I got my own second-hand copy and it was from this green-bound book with a golden dragonfly impressed on the front that I learnt the rudiments of the life histories of the important groups of insects that pass their immature stages in fresh water.

2

A Boy in the Forest

When I was eleven I passed the examination to enter a grammar school. This was King Edward VI School, Aston, in my native parish, but about a mile away from home. I did not gain a scholarship and my parents had to find eight guineas each term and a substantial amount for books. I must have been on the short list for a scholarship because I attended an oral examination which was conducted by the second master. This consisted of a first lesson in Latin involving such sentences as, *Scutum latum est* and *Saggittae acutae sunt* and we were afterwards examined on it.

One's impressions of school—carried on into later life—are necessarily fragmentary. Incidents stand out, at least some of them, and not always the expected ones. I cannot remember the first time I played in the first XV at rugger and yet it seemed terribly important at the time. I have just got a hazy recollection of how important it was and no more. I was never a regular member of the first XV and yet how jealous of Charles Tidmarsh I was—he always had a place, quite unjustifiably I thought. But, at school, games never loomed large for me. They took up a very much larger part of my time when I had left school and before I married.

I have always considered myself fortunate in my school life, although I sometimes criticize important omissions from the school syllabus and lack of encouragement for my then unconventional pursuits. I think, without exception, that my masters were extremely competent. Strangely enough, apart from idiosyncrasies of some of them, it is recollections of punishment that stand out in my mind. A schoolboy's sense of justice is, I think, a very accurate one. When I was punished I knew very well whether I deserved it or not. Unfortunately some masters keep above their scholars' heads the fear not of just punishment but of unjust, such as the various methods of collective punishment. But I knew I had been justly punished when on that hot summer's afternoon I fell asleep during a physics lecture—and it is one of those outstanding incidents of my school years. So very vivid are my impressions of that afternoon forty years ago that when I think about it I can live through the experience again. The hot dusty room, the gritty, yet polished teak tops of the laboratory benches and the little clusters of gas tap nozzles. The narrow windows looked out over the blue slate roofs which ran in a monotonous jungle from the breweries and vinegar factory to the Birchfield Road without the relieving luxury of a tree. When the physics master came into the room he quickly got under way with one foot on the dais of the demonstration bench and one on the laboratory floor, but with both hands in the trouser pockets of his bulky grey tweeds. 'A calorie is the amount of heat required to raise the temperature of one gram of water through one degree centigrade', but then, slightly further away, 'A calorie is the amount of heat required to raise the temperature of one gram of water through one degree centigrade'. Then the rows of still little boys precariously balanced on high stools gave way to a sedgy meadow with a swiftly moving brook, and it was ice-cold when a hand dipped in. It was still and quiet, the grittiness had gone. Things could take their time, nothing was important. I slipped my hand splashlessly into the cavity under the bank made by some old twisting roots. I touched the trout, then unhurriedly moved my questing fingers towards the gills ready for the sudden grab whilst I held my breath. But the scene changed with startling suddenness. The stream disappeared and a great moon-shaped face loomed up in front of me, and rows of small faces turned, watching the awful spectacle of a boy who had fallen asleep. 'The question is—what is a calorie?—and this question is reserved for Master Hickin.' No reply was made, and the taunts of detention did not sting; after all, I had failed to keep awake. The exactitudes of the physical sciences have never appealed to me but the definition of the unit of heat has remained firmly fixed in my memory.

Tubby had put in a rather shoddy essay. He was not one of the brilliant ones, but usually he kept out of serious trouble. I seem to remember that it was the roughly torn edge of the sheets on which he had written his sketchy remarks that roused our master's ire.

'Come here, Hayden. How old are you?'

'Fifteen, thir.' Tubby blenched as he lisped it out.

'You know I never cane a boy when he's sixteen, I like to give him a sense of dignity, perhaps punish him in other ways if he deserves it,' our master mouthed as he fingered his cane, whilst his staring eyes caught those of the luckless Tubby, who had now presented himself before his master. The master's hand shot out, caught Tubby's jacket on the shoulder and pulled his head down until he had him trapped between his knees. Tubby's buttocks were making the closest possible contact with the thin worsted of his breeches and the cane whirred and sang six times before the victim was released to bravely retrace his steps to his seat.

The sacrifice seemed to be over. We seemed a little more free after a period of uneasiness, but the master had not finished. 'Hickin—how old are you? You are getting a big fellow.'

'Fifteen, sir,' I answered and smiled warily as I untwisted my long legs under the cramping desk.

'Yes, I know,' he said, 'but when will you be sixteen?'

'Tomorrow, sir!' I replied, again smiling, this time with the pleasurable anticipation of a birthday on the morrow.

'Then I shan't have the pleasure of caning you again, Hickin.'

'No, sir,' I said.

Then his tone changed. 'Well, come out here and we'll rectify the omission, I know you wouldn't deny an old man a little pleasure.'

So I bent down and received six strokes. I hope they gave him pleasure, for they hurt me like hell.

My handwriting has always been bad but only once has it got me into serious trouble. Often, I tried to improve it but my writing usually got by as the subject-matter of my school written work was mostly satisfactory. The Head seldom came into the form room and then it was usually to make enquiry about some detail of administration such as, 'Has any boy changed his home address?' Just by the way, this question once brought the astonishing reply from a boy who had not heard correctly, 'Yes, sir, I am now wearing a waistcoat.' It seems obvious to me now that the Head must have had a report from an inspector commenting unfavourably on the standard of handwriting, because he came into the form room one afternoon and made us cover our desks with our open notebooks. He was edgy and nervous and spoke sharply to several boys, myself included. Joe, as we affectionately called him, was a big shaggy man, but always tidy, and his hair only slightly longer than a crop. I was one of those to whom a copy-book was given with orders to complete it by the following morning. Judging by the subsequent attendance, the whole school must have been visited by Joe, and three or four of the worst caligraphists in each form singled out for special attention.

Joe was quite angry when he saw my poor effort and asked whether I had anything to say before he gave me a caning. I spoke up. I said that I thought it 'unjust' to be caned for handwriting when my position in the class was invariably among the top four places and that I was top of the form in science. I think this touched Joe, but he would not go back. 'There's a lot of injustice in this world, Hickin, and you may as well just have a taste of it sooner rather than later.' With that he took me out of his room and caned me hard on both hands in front of Fleetwood Walker's portrait of Captain Phillips, V.C.

What extraordinary good fortune I had in learning how to draw! On the day I commenced at Aston Grammar School a new art master also started. He was soon given the nickname of 'Twisty Bill', but I've not the faintest idea why. His hair was brushed straight back and he always wore a stiff fly-away collar, and he had prominent nostrils. As a master he was eccentric and had the habit of flying off the handle at boys whose work was 'smelly' or 'rubbishy', and I vividly remember his action when a ruler was found in the classroom. He just opened the window, threw it out and went on with the job in hand without uttering a word, whilst we breathlessly awaited the sound of the ruler coming to rest with a clatter on the pavement two floors below.

I was very keen on drawing and had a little natural aptitude and it was not long before Fleetwood Walker was spending very much more than my share of his time teaching me a lot about drawing and the use of water-colours. This subsequently turned out to be a great asset when I became a student of botany and zoology and when, after graduating, I took up the task of working out and describing the immature stages of the caddis flies. I illustrated this long series of papers very profusely with line-drawings.

Bernard Fleetwood Walker is now the well-known portrait painter and royal academician, and for many years his outstanding portraits have graced the Royal Academy, but the trouble he took with a small boy, teaching him how to draw a curved line, has given great pleasure to that boy and helped to serve the cause of biological sciences.

Those of us in the upper fifth who were leaving, lined up outside Joe's study for a few words with the old man. Matric results, of course, were not yet out. When it came to my turn Joe asked me what I wanted to do and I replied that I thought I would try to get into the chemical laboratory of one of the local industrial firms and work at the Technical College at night for a B.Sc. degree of the University of London. I thought that if I got this I might ultimately get a job in biology. Joe dropped his chin a bit and looked at me over his glasses. 'You know, Hickin,' he said, 'you have to be one of the brilliant ones to accomplish that and be gifted with extraordinary doggedness, taking the greatest care that a breakdown in health doesn't ensue.' About getting into biology he was less than sanguine. 'There are only a few jobs in the whole of the country, mostly in universities, and mostly held by women,' Joe said. If I wanted to earn a living in science Joe thought it would have to be in chemistry or physics; biology—not a hope. I did not learn for many years afterwards that his own university degrees were external ones.

I was not very good at Latin but there was one aspect of the language in which I excelled: the vocabulary. This was on account of my knowledge of the generic names of animals and plants. We had the system of standing around the classroom whilst the master fired rapid questions at every one in turn. If you did not know the answer the next lower down was asked until someone knew, then he went up so many places. On vocabulary I usually went to the top of the class but when grammar came round I gradually lost ground until I was near the bottom. I had a number of tussles with masters. On one occasion I was asked in a Latin lesson for the plural of 'two' and I hazarded a guess and I think I said duocimi or something like that. 'You fool, Hickin,' my master said. 'There's no plural of two.'

I said, 'Surely there is, sir.'

He lost his temper a little and said, 'All right, give me an example,' and like a flash I was able to say, 'The soldiers marched down the street in twos.'

When in the lower fifth form at Aston and just about fourteen years old, I made my way to the Reference Library at Ratcliffe Place. I had been given an essay to write on Marcus Aurelius, and two or three evenings were spent getting information from the *Encyclopaedia Britannica* and other reference books. This set the pattern for my essay writing and I was a regular attender for about twenty years at this amazing library, the like of which I have never seen since. I do not know how many volumes there were staring down at me from the acres of wall space, but books were obtained for you in an incredible short space of time by filling in the title, author and your name and address on a tiny piece of paper. In addition there was an 'open access' section for the commonly used reference books.

At Aston cribbing, or, quite bluntly, cheating, took place on a very wide front; in the lower fifth form the work done by a boy to give his masters the idea that he had done certain work (when indeed he had not done so) probably exceeded the work originally required. Some boys with keen, agile, if somewhat distorted minds, did surprisingly well. I never did well at it and got into serious trouble on several occasions when I got involved with it. I lost a Thursday afternoon holiday by allowing my answer to a question in chemistry to be copied and distributed to the rest of the class. I was fairly good at the problems set in quantitative chemistry and everyone in the form relied on me to produce the correct answer. On this occasion I made a rather silly mistake which everyone religiously copied down. It was easily found out and I was the only member of the form to be punished, rather unjustly I thought as the senior chemistry master lectured me, making me squirm in my seat as he went on about my competence in chemistry and my lack of elementary skill in cheating. I did answer back by intimating that I did not think that I was the one who had cheated, but a withering glance silenced me.

The Easter rides which the smiling and tubby junior physics master ran were an important part of school life. They were amazingly successful. Four or five times each Easter holiday we met at the school with our bicycles and then cycled to some place about twenty miles away. Stratford-upon-Avon, Kenilworth Castle and Bewdley we went to every year. When we got there we had a picnic lunch and sometimes funds ran to tea in a teashop. We usually arrived home in the early evening.

My impression was that the rides took place on bitterly cold days. These later years I have travelled about the British Isles extensively but have never seen the spectacle of thirty vividly capped schoolboys cycling along in a compact body with a master in the centre like the queen in a swarm of bees. Actually he never taught physics to me, but I belonged to his science club and he went to endless trouble to show us interesting scientific phenomena. One such was the passage of electrical discharge through rarefied gases and how one could bend the passage of the electrons with a magnet and show where they were by means of fluorescent layers. We did not know that this was the television tube of the future.

There is no doubt that one of the important sources of encouragement in my study of natural history was the Birmingham Museum and Art Gallery. The impressive flights of white marble stairs leading ever upwards was in sharp contrast with the Technical College entrance. Admission, of course, was free, but one had to click through a turnstile, thereafter the gleaming marble and brass handrail led one past Venus Aphrodite, Epstein busts and Burne-Jones tapestries to the natural history collection. The Feeney collections of nesting British birds must be the finest of such collections, surely the peak of the taxidermist's art. It never failed to excite my interest as each case was an ecological study. I could make a complete list of the exhibits in the museum, from the skeleton of the Irish elk and a rhinoceros to a large wall case showing the probable evolution of the butterflies and moths.

Then, one Saturday morning when I was fourteen, I got an introduction to the curator, and was shown into a little dusty room full of insect cabinets and storeboxes. I suppose I looked like any other of the dozens of small boys ushered into his room who collected butterflies. Anyway, the cold-water tap was turned on and the kindly and bespectacled old gentleman very firmly but nicely detailed the long hours of study required to obtain qualifications necessary for museum appointments and the great competition for jobs and the low salaries. This failed to daunt me and certainly had no effect on my insect collecting. During spring, summer and autumn the house was literally full of boxes and jars containing such caterpillars as the buff-tip, lime hawk and the drinker, and anything that would contain more than a gallon of water was pressed into service as an aquarium.

The year before I went to the grammar school in 1920 we went as a family on holiday to Bewdley, a little town on the river Severn, and we continued to go until I left school in 1926. We stayed at a little cottage-cum-teashop guest-house, presided over by a large, good-natured landlady and her daughters. We always had a very happy time there, often accompanied by other members of my mother's family. We got to know the countryside very well; Blackstone Rock, Devil's Spadeful, Habberley Valley, and we ventured into Wyre Forest. We accomplished the latter walk by the same route taken by many naturalists, geologists and writers for well over a hundred years, as I know now from their books. One first makes for the spot where Dowles Brook flows into the river Severn. This is about a mile north of Bewdley on the west bank and can be approached along the path on the river bank. Dowles Brook, which cuts the forest into two by flowing west to east, is followed sometimes on one side and sometimes on the other. The tips of two tree-fringed meadowlets run steeply down to Dowles at Rudds Bridge. A little mill was formerly sited here, but so speedily does the forest cover with mould and moss where folk have failed to maintain their little plots that a search is required to find it. There is a ford here also where timber from Brandwood Coppice was brought across the stone-bottomed brook and hauled with many sweating and straining horses up the opposite track. This was so steep that it had to be surfaced with oak branches for the steel-shod wooden wheels to grip. This oaken road is as sound as steel today, but traversed only by tractor tyres not three times in a year. This was the road that climbed to Wyre Forest railway station. One takes a train back to Bewdley, the single railway line also bisecting the forest. The station after a hundred years of service is now closed and no passenger trains whistle at Ford Lane crossing.

The Forest of Wyre was formerly divided into that part near to Bewdley called Bewdley Forest and that part further away called Far Forest. Inhabitants of the countryside around merely refer to 'the forest' and they say they live 'on the forest'. Far Forest has come to mean a relatively small area including a small village five miles from Bewdley on the road to Cleobury Mortimer just as it plunges down to Lem Brook and crosses into Salop. This is the Far Forest of Francis Brett Young's novel of that title. The Far Forest school is a landmark for many miles around and when the skyline can be seen to the south-west from the forest through the oak branches so can be made out the little spire under which is the text, 'feed my lambs'.

I first camped in the Wyre Forest when I was eleven years old. A school friend's father, who was a captain in the Royal Warwickshire Regiment, had arranged for two bell-tents to be left up after a Territorial Army camp, a few fields away from the Horse and Jockey Inn. Lem Brook flowed ice-cold and clear, only a hundred yards away. We did not wander far and it was wet—so wet one night that we slept in a barn on the remnants of old hay. I have never slept in such a barn since and the memories of the irritation I suffered from insect bites and bits of short hay sticking in my neck will deter me from ever repeating the experience.

It was during this camp that a minor event occurred which had an important influence on me. I had made a catapult, a robust affair, and I had some idea that I should stalk and finally kill a rabbit for our dinner. On the evening of our arrival at the camp I commenced to test out this instrument, using a glass 'marble', of the type then used as a valve in a lemonade bottle, as ammunition. Pulling the elastic back as far as my strength would allow, I let the missile fly towards a group of swallows that were whirling about high in the sky almost out of sight. To my dismay a swallow came tumbling down and fell dead at my feet. Since that day I have never directed a weapon at a bird and have abhorred the killing of animals for sport. I began to ponder on the exact circumstances in which a man is justified in taking life. Later, when I killed animals for teaching purposes and, later still, killed insects on a vast scale, the subject of killing worried me more and more. In the end, however, I made my own rules of behaviour and came to terms with the problem.

In succeeding years, but whilst still at school, I began to know the forest and the plants and insects it contained. During the summer months I was almost irresistibly drawn, especially during July when everything in the forest seemed to reach its climax.

July was, of course, the month of the silver-washed fritillary butterfly. Along the Dowles they literally fought for the thistle heads and when the males had freshly emerged we could follow their flight over the oaks a quarter of a mile away. The females were larger, yellower and greener, and when I was fourteen years old I watched them laying their eggs on the lichen-encrusted bark of the oak trunks. This thrilled me intensely and I can feel my heart stopping yet again as, like a statue, I hardly dared to breathe lest I should frighten the magnificently beautiful silver-washed fritillary as she came towards me fluttering around the oak trunks that bordered the path on the brook's bank. Then she rested for a few brief seconds whilst she curved her body to place an egg amongst the encrusted lichen on a trunk only two feet away from my eyes. Then she went on to the next oak where my eyes followed her, but it took me two or three minutes to find the eggs again.

The thrill of discovering this egg-laying habit of the silver-washed fritillary exhilarated me, but on searching through the butterfly life

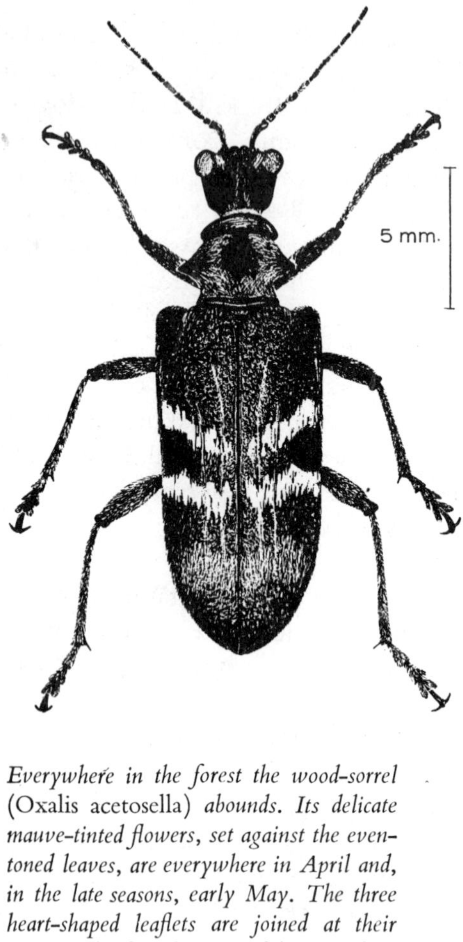

When the silver-washed fritillary butterflies emerged during the first ten days of July I would walk along Dowles past the island to their favourite haunts. In the bright sun the large longhorn beetle Rhagium mordax *would fly encircling me and then alight on my sweater and their sharp tarsal claws would get tangled up in the wool. I still have some specimens caught when I was twelve which I wrongly identified as its near relative* Rhagium inquisitor. *I relied too much on the description of the specific name.*

Everywhere in the forest the wood-sorrel (Oxalis acetosella) *abounds. Its delicate mauve-tinted flowers, set against the even-toned leaves, are everywhere in April and, in the late seasons, early May. The three heart-shaped leaflets are joined at their apices and reflect the state of dryness of the air by opening and closing. When damp, the leaflets are extended flat, and lie in the same plane, but when conditions are dry the leaflets droop and in doing so the lower surfaces are pressed together as each leaflet is reflexed down the centre. In this way the stomata are protected from water loss. The origin of the Irish Shamrock emblem was undoubtedly the leaf of the wood-sorrel, but today the leaf of the white clover* (Trifolium repens) *takes its place although of relatively recent introduction into Ireland.*

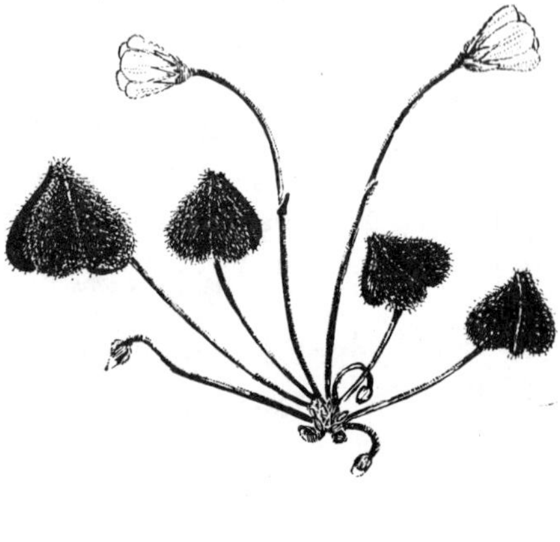

histories in Frohawk's book I found he mentioned it. When the minute caterpillars hatch out during August they stay amongst the lichen on the oak bark and do not descend the tree until spring to search out the violet plants, the leaves of which are their only foodstuff.

The other two large British fritillary butterflies were also found in Wyre. The 'highbrown' was very common but I only saw a few examples of the 'dark green'. The spiky caterpillars share the great abundance of violet leaves with those of the 'silver-washed'. Both species of the pearl-bordered fritillary were very common and appeared when the forest paths were colourful with ragged robin and bugle. The greasy fritillary became very common along Dowles for a few years but was exterminated by collectors. After hibernation the compact mass of black larvae in their communal web on the scabious plants is easily identifiable and large numbers of these broods were taken back to Birmingham.

The bicycle was my means of transport from the age of eleven until the end of the Second World War. Any other transport was luxury indeed. But when I was a boy and a young man there was no difficulty about getting from one place to another, one just threw a leg over a bike and pushed oneself there. Several times I cycled over 170 miles between dawn and dusk. From home to Bewdley took one hour and fifteen minutes but slightly longer coming back on account of Mucklows Hill, a formidable obstacle to the cyclist.

From Birchfield Road there were several ways of getting on to the Hagley Road and thence to Bewdley, just over twenty miles from home. Usually we took the long grinding ride up to St. Michael's Church, then down and up to Winson Green Prison, which was skirted with some foreboding, but when we came to the City Road the feeling of leaving the city behind began to come over us. There was a fleeting glimpse of some fields before the main Hagley Road was reached, at the King's Head. The road was very hilly and the Black Country actually touched the road at several points. In addition, the road made short traverses through the steep little streets of Halesowen, giving wide views of smoking cinder banks, and groups of chimney stacks pierced the blue-grey smoke hanging over Dudley and Brierley Hill. Then quite suddenly rural Worcestershire was reached with the breathtaking and glorious free-wheel down Hagley Hill; only the run-down off the Malverns towards Worcester ever equalled this.

At Blakedown we sometimes turned right to go through Franche and Wolverly as a diversion from Kidderminster, but when we got to the top of the hill out of Kidderminster we could not only see the forest but the Brown Clee Hills behind. So far the road only was familiar to me, but now the leaves, fields and woods I knew.

Although the Button Oak road ran parallel with the river as far as the Dowles Bridge, we always preferred the switchback rough ride along the bank of the Severn. One or two pike fishers were usually busy casting their spoon baits or hooked fish into the swirling water. Just past the derelict brick kilns the little red brick church of Dowles came into view, partly hidden by trees. When I was quite young the figure of the Rev. John Burton was quite a familiar sight. He was dressed in the fashion of some fifty years before with a biretta on his head and a coat with tails. It had been black but now was as green as the fields and the young oak leaves of his beloved Dowles. It was not until years afterwards that I learned his identity and found a copy of his book on the history of Bewdley in the Charing Cross Road. This scholarly gentleman, finding me near his church one day, invited me inside to see the silver chalice and plate. He took them from a little safe and explained the markings, and then, taking me outside, showed me where a large chemical works had once stood but now was gone without a trace. How strange that his own church has suffered the same fate, only a few gravestones marking the site today.

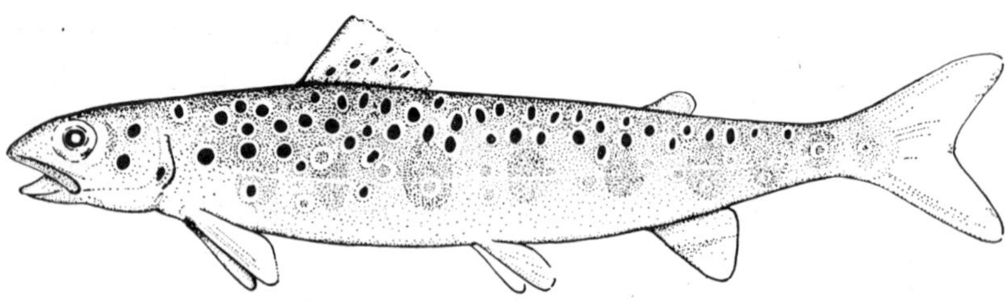

Dowles trout (Salmo trutta) *are lean and sharp. The coldness of the water and the lack of aquatic vegetation probably accounts for their slimness. Mostly, you see them first when they have already seen you as they flash from their feeding position in mid-stream at the tail of a ripple where the water gushes between two rock slabs. They take shelter amongst the washed-out clusters of alder roots. Eight-inch fish can even be found in cattle drinking puddles in the little streamlets running into Dowles but still half a mile from it, and the little side-blotched one-inch-long trout can be found in the tiny rills all about the forest.*

As all readers of Richard South's Moths of the British Isles *will know, the rare Kentish glory moth* (Endromis versicolor) *is not uncommon in Wyre Forest. Elsewhere in the Midlands it was at one time found in Herefordshire. Its distribution in the British Isles is discontinuous, being found in the north of Scotland and a few scattered localities in the south of England, but mainly in Sussex. It is a large handsome hairy moth mottled with browns and white with two dark bars across the forewings. The male is small and much darker than the female, its hind wings being almost brick red.*

The caterpillars of the Kentish glory (Endromis versicolor) *feed on the leaves of the silver birch with which the coppices of oaks are mixed. In their early stages, to some extent, they are gregarious; the eggs having been laid in a batch at the tip of a twig, the young larvae keep together until their large appetites start them wandering. When not feeding they assume a remarkable attitude which makes them very difficult to detect so like are they to a birch leaf. Furthermore, if they are touched or their fore part deflected with a twig they become stiffened, paralysed and, indeed, behave like a leaf.*

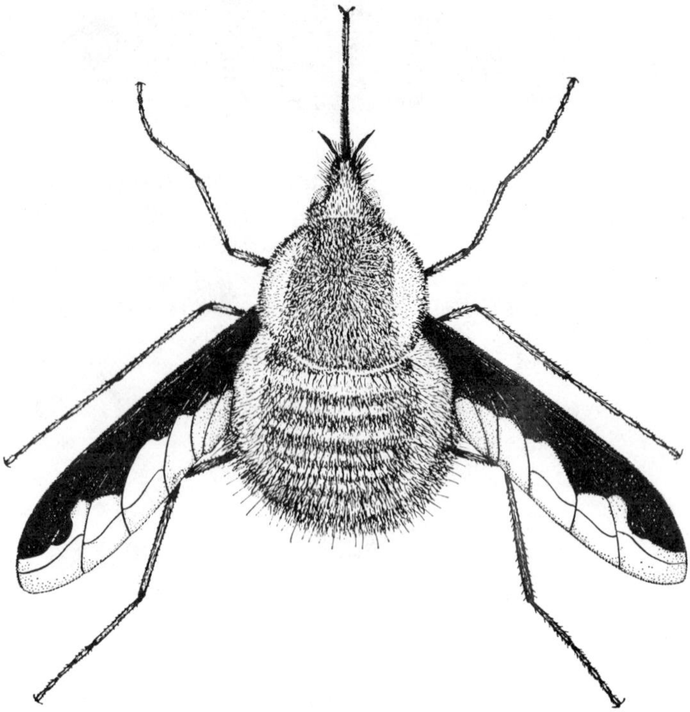

The great bee fly (Bombylius major) *is usually seen hovering with invisible wings, but when at rest it is observed that the front part of the swept-back wings are smoky black.*

As well as looking like a bee, this fly is a parasite of one of the andrenid mining bees. The female Bombylius *scatters her eggs in flight in the vicinity of the tunnel in the sand bank which leads to the bee's brood and the minute and agile larva makes its way down the tunnel and gets into a bee grub's cell. Growth of the uninvited guest is slow at first as it consumes some of the nectar store which had been collected for the bee grub, but when the latter is nearly full-grown the* Bombylius *assumes a more maggot-like appearance and turning its attention to its host fairly rapidly sucks it dry.*

The general similarity which often exists between host and parasite in the insect world has intrigued entomologists for over two hundred years.

Dowles Bridge of latticed girders brought the fussing little two-coach train high up over the river from Bewdley on its way to Cleobury Mortimer and Tenbury Wells. The Dowles Brook flowed out of the forest at this place. Dowles was deep here, but the Severn was shallow. A short path then brought us to the Bridgnorth road, and crossing it under the railway bridge a straight steep path took us down to the brook and to a sight which has always gladdened my heart: the view of Dowles Manor where the black and white elevation and clipped yew trees and little stone walls gave the perfect feeling of rest and content. It always displayed dignity and charm. It was in the possession of the Steele-Eliots—a family distinguished in science. I passed down the path several hundred times from 1921 and every time I stopped by the group of larch trees and peered through the branches to the house. No other house has ever moved me thus.

I suppose Dowles Manor was at the root of my own obsession to live in an old house. As Emma says, I had to get it out of my system, and Home Farm, Fetcham, is, of course, by contrast with Dowles Manor, quite a modern building, having been erected in 1667. I remember Emma giving the perfect rejoinder to a friend who was expressing concern at our living in a very old house. 'I wonder how many people have died in this house,' he said, to which Emma replied, 'Roughly equal to the number of babies that have been born in it, I suppose.'

It must be explained that the fascination of the forest for me lay not only in the quiet rides or the brook bank, but also the little farms that lay on the half-forgotten tracks and lanes leading into the oak and birch woods. All around the forest was charming country and still is to this day. In mid-April, viewing the forest from any high point, it resembled a gigantic grey petticoat edged round with pure white lace of the flowering cherry orchards, and the gean or wild cherry here and there extending the frilly whiteness into the forest itself.

3

Forest Camping Days

THE great majority of boys at grammar schools left at the age of sixteen, and I was no exception. There seemed to be no chance for me to go into the university. Before leaving school, as mentioned earlier, I had the notion that I would try to get into one of the chemical or metallurgical laboratories that abounded in the great industrial belt around Birmingham. At that time, 1926, it must be remembered there was great unemployment, it was the year of the general strike and most companies were content to merely exercise a holding action. I had taken the school certificate of the Northern Universities Joint Board and had gained sufficient 'credits' to obtain the matriculation certificate, but my first round of letters to industrial firms applying for a job did not get me anywhere. However, I answered an advertisement for a laboratory assistant at the United Yeast Company and I got the job. This was interesting work, helping with the analysis of various foodstuffs sold in the confectionary trade, mainly in connection with the detection of poisonous preservative materials in imported food, such as arsenic and boric acid in Chinese egg-yolk which was imported in forty-gallon barrels. I also tested the rising power of many brands of yeast and this was done by baking standard loaves. It struck me as rather unusual that one could eat the experiment afterwards.

When an opportunity came my way to apply for a laboratory assistant's job at the Dunlop Rubber Company's factory at Erdington I took it and I got the job. I started work on 1st January 1927 at sixteen years of age and my wage was twenty-five shillings weekly. The laboratory occupied a large wooden hut with a hemispherical roof, a relic of the First World War. It was divided into 'bays' running off a central aisle, with offices, stores and weighing room at one end. We were presided over by a short, very dark, balding and nervously energetic man, Dr. Douglas Frank Twiss, the chief chemist, and he provided the element of discipline and decision which I required. He was a world authority on the vulcanization of rubber, the chemical bonding of natural rubber with sulphur, which so improved its properties that the pneumatic tyre, and thus modern motor transport, was the natural development. He had written a great monograph on the element sulphur and we were all very proud of him. Dougy, as we affectionately called him amongst ourselves, had been a lecturer in the department of chemistry at the Birmingham Central Technical College, and he had strong and determined views on how young men should comport themselves.

On appointment we were given to understand that we were expected to work for an external degree of the University of London and to attend evening classes regularly to that end. We were allowed to leave at five o'clock on those evenings we went to 'night school', instead of the usual five-thirty. This allowed ample time for the cycle ride into the city and a cup of tea in the refectory before class, if funds allowed, which was not very often. I was expert, however, in extracting a bread and butter and jam sandwich from the saddlebag of my racing cycle and eating it going along.

After class there was the long cycle ride home over the stone setts. When the roads were wet and slippery, which was often, crossing and recrossing the tram-lines was a nightmare. The light weight of the cycle of that time, with inch-and-a-quarter rims and calliper brakes, was not an efficient instrument for dealing with wet tram-lines. If the angle of crossing was too narrow the front wheel was jerked violently out of one's grasp and you were sliding over the muddy road before you realized what had happened, and the next fifteen minutes were occupied straightening the pedal cranks. One such dark, wet and misty night is remembered well. I was going fast down Corporation Street when suddenly I met the lines at the wrong angle when coming down Steelhouse Lane, appropriately enough just outside the General Hospital, and somehow I fell on top of my cycle. Dazed, I dragged my twisted cycle on to the pavement, sat there for ten minutes before I started the job I knew so well—making a bicycle rideable.

Each bay in the laboratory was in charge of a senior who specialized in one or more aspects of the work and there was an annual move round. My first work was the estimation of free sulphur in the reinforcing red mineral pigment antimony sulphide, known as golden antimony and which came from Japan. It was the reason why motor tubes of that time were red. The free sulphur content was determined by extracting in a soxhlet condenser with the solvent carbon disulphide. At the end of four hours' extraction the solvent was evaporated off, and the sulphur, which looked like a yellow smear, weighed in a flask. From this the amount of sulphur was calculated which was required to bring up the percentage of free sulphur to a predetermined figure. I can remember it now—it was 17·5. My day's work was about twenty estimations. Then, about the middle of the afternoon, armed with cask numbers and the weights of sulphur to be added, I went to the powder floor in the main factory where the various compounding ingredients that were to go into the various types of rubber mixes were ready in opened casks. The required amount of sulphur was weighed out and put into its respective cask with a large 'passed' notice placed on it. This was a filthy, dirty and most disagreeable job. On the floor, as, indeed, lying on every object, was a layer of a peculiar dusty powder, made up of all the powder ingredients of rubber compounds, with carbon black and zinc oxide predominating, but laced with some foulest-smelling chemical compounds, such as zinc isopropyl xanthate and mercaptobenz thiazole. I remember their names so well. Actually, every cask, drum and keg of chemical and every consignment of everything purchased by the company, such as soap and petrol, was analysed in the laboratory for fitness for purpose. Today, of course, such testing is the responsibility of the vendor.

During the hot dusty summer afternoon there was something vaguely satisfying about the weighing room where twenty oertling chemical balances were set out on cement slabs around the room. It was quiet and anyone coming into the room, with his desiccator of crucibles and flasks to be weighed, knocked on the door and allowed sufficient time for all the sliding doors to the balances to be shut before he entered. We weighed our crucibles, or whatever it was, to four places of decimals of a gram by what is now an obsolete method. We placed the rider along the beam with great precision and dexterity and we left the weighing room with satisfaction, knowing that from our determinations of mass we could calculate all the data required of us. Strangely, the weighing room was the location of fits of temper. I once had a tin of zinc oxide flung at me as a result of what I thought an innocent remark. This resulted in an electric afternoon during which I mopped the perspiration off my face many times.

I did a stint in the oils, fats and waxes bay under the supervision of a senior and carried out a very long series of the usual tests of these materials: flash points, viscosity, iodine value, saponifiable matter, etc. I think I could still carry out these tests if I had a glance at the laboratory methods book first. A very large industrial factory like Dunlop's at Castle Bromwich used a lot of steam for the vulcanization process and also to make electricity, and 200 tons of coal was burned each day to do this. Every truck of coal was analysed for moisture content and ash and its calorific value determined in a Mahler-Cooke bomb, and when I moved on I had a fair knowledge of the characteristics of the products of the main British collieries.

One story I have repeated very many times during the years which have passed and which I would like to tell here. During a period when the company was not doing very well financially, an expert in business efficiency, as it was then called, was brought in from America. He spent several months at the factory and when he got to the chemical laboratory he interviewed everyone and asked them about their work. One day he met one of my colleagues and on being introduced asked him what he did. He replied something like this: 'Well, you know, sir, as soon as I arrive in the morning people start interrupting me and I get a large number of enquiries about nothing in particular. Honestly, when I put my coat on at night I don't know what I've done.' This surely must be the way *not* to have answered that particular question, and, indeed, it got him into trouble with Dougy, more, I think, because it was the wrong sort of answer, rather than because he did nothing, which was not true.

Of course, Dougy had a number of detractors and I remember one official saying to him: 'What are you going to do with all the young men, Dr. Twiss? We shall not be able to employ them when they are all qualified scientists.' How wrong he was! The company now employs many more qualified scientists than it trained, although the men that Dougy trained are now all over the world, inside and out of the rubber industry. Many have attained outstanding appointments not only in industry but outside it, and I consider myself very lucky to have grown up with them under Dougy's discipline.

The railway siding where our daily twenty trucks came in was adjacent to Bromford Bridge racecourse and we often had a grand-stand view of the racing from the top of a coal truck. Like most industrial sites around Birmingham it was a dreary scene, composed mostly of reserve coal dumps, ground made up with ash and deep ditches carrying polluted water. But the rosebay willow herb grew by the acre and the large larvae of the elephant hawk moth were abundant feeding on it. The skylark and plover often made their nests or scrapes on the ash, and I once found a swallow nesting in an underground sewer six feet below ground level. Now, when I find myself travelling along the Chester Road where it flanks the old flats I knew so well as a boy, I think of the auroch's horn that was found during the widening of the bridge over the stinking river, and I find myself transported back in time and hear the grunts and splashing of our ancestral cattle as they wallowed in the reedy beds before the wheel was invented.

I spent several years in the physical testing laboratory at Dunlop's, carrying out a seemingly endless series of measurements of various physical properties of rubber compounds. The experimental rubber compounds were vulcanized under a series of conditions usually for differing times at a certain steam pressure, for example: 20, 40, 60 and 80 minutes at 40 lb. per square inch. Rings were cut from slabs used for tensile strength and permanent set testing. Blocks were used in several testing jigs for estimation of several properties related to elasticity. The 'pendulum' was one of these and required two operators, one wearing headphones listening for clicks which signified contact. The swinging of the pendulum and measuring the extent of its rebound required much manual dexterity, but after attaining competence in the manipulative techniques of these tests they became deadly dull. One machine I especially remember, because I had spent so much of my youth on it, was a machine for testing the abrasion resistance of rubber compounds. Small wheels were vulcanized on to steel cores and were run against standard carborundum wheels under certain fixed conditions of angle and slip, and we weighed the amount of rubber which had been removed during the test. I made a contribution to this test by devising a method of picking up minute pieces of rubber which had been cut off by abrasion by another wheel, the sticky idler, which I called it because it was slightly stickier than the rubber compound. This was at a time when tyre-tread compounds were themselves rather sticky and the tiny pieces of rubber cut off during abrasion tended to stick together and give false readings of abrasion resistance. I spent some time in charge of the ageing tests on rubber compounds. These were to evaluate the probable life of the compound by subjecting it for various times to an increased temperature and to an increased pressure of oxygen. But I was not happy with physical testing and whilst watching the dials of the testing machines I would be thinking of the leafy forest and of what species of moths and beetles would be abounding.

At school I was a rugby football enthusiast, although I never played regularly in the first XV, only managing about six times, although I felt that I deserved a regular place. But our house matches were keenly contested. Although we had no boarding houses, the school was divided into four houses, each named after a master, and I was in Mr. Ross's house. When I left school at sixteen I joined the Aston Old Edwardian Rugby Football Club, which then fielded six teams, and I started to play in the 'A', which was, in fact, the third team. Before I left school I weighed 13 stone 12 lb., but I was fairly agile.

The following season I started to play in the second team, but before the end of the season I was in the first XV, and played regularly during the following eight years. I loved every minute of it! A long succession of players' names can run off my lips: and they somehow sound like poetry. For two seasons I packed in the second row of the scrum and thereafter hooked in the middle of the front row. I enjoyed the vigour and vitality and the striving and the expenditure of energy, without thought of mud or slush and not counting the cost in terms of bruises, cuts, stiffness and other injuries.

During my last eight seasons I never missed a match for any reason whatsoever, and this included Easter tours, which were very popular, playing three games in four days. If I went on to the field with a snorting cold I had lost it by the time I had emerged from the steaming communal bath, although perhaps I had passed it on to other members of the team. It never failed to thrill me to put on the jersey and stockings of white, green and red hoops, but I was never conscious of providing a spectacle for others. My last match was against Weston-super-Mare, when I got a heel with a lot of boot on it in my mouth, and lost all my teeth except four. Since that Saturday in February 1936 I have only seen one match and I have never taken the slightest interest in the game. But when I played it was very important to me and looking back I feel that its very classlessness had an attraction for me.

Cricket I never played very much for the all-important reason that it is played in the summer when I spend every leisure-time minute in pursuit of natural history along the lanes or in the woods.

During the chem lab days I usually had two or three and sometimes many more companions during my weekend or holiday camps in the forest. Usually they were young colleagues from the lab, but in addition one or other of my friends and my brother's friends from our grammar-school days accompanied us.

We did not leave the laboratory until noon on Saturday, but all our weekend stuff was packed in old army packs ready to be shouldered as we cocked our legs over our bikes and got off from the black ash-strewn cycle sheds at Dunlop's. The old army pack was a remarkable object and for years and years it served me as a container for anything I wanted to carry. A relic of the First World War, it was rectangular and was composed of very thick khaki canvas. The two fastenings and the two shoulder straps were virtually indestructible and it was waterproof. At a later date it became fashionable to cycle with a saddlebag (a rather elaborately shaped container slung under the saddle), although I hung on to the idea of carrying everything on my back for sheer simplicity's sake for many years; whether it was my collecting and camping equipment *en route* for the forest or a load of books *en route* for the Tech.

When I had my insect-collecting and camping equipment on my back the Wyre Forest was not my only Mecca. When talking to fellow lepidopterists I always had my ears pricked up for mention of localities where the rarer moths and other insects could be found. I once went to Wilnecote in the Staffordshire coalfield instead of Wilmcote in the Forest of Arden.

But Cranham stands out as a great discovery. This little Cotswold village, with its beechwoods, and high-striding common, I first visited at the age of sixteen. The only directions given to us were that we turned left just before we got to the Royal William on the Cheltenham–Stroud road. Dusk was falling as we reached the critical landmark and when we actually saw the signpost, 'Cranham one mile' we took the wrong fork in our excitement. After we thought we had gone a mile we were still going through beechwoods, now very dark. Then we realized we were above the village and picking up our cycles we carried them down the beechy banks until we came to a clear area by a brook. It did not take many minutes to erect our little tent even in the dark, then we made our way with our nets to the high bank of the common which we could see against the western sky. We caught many species that we had never caught before as we were on chalk for the first time. Quite accidentally we found the day-flying green forester moths in our nets, so that when we eventually crawled into our blankets we knew that we should see extraordinary butterfly and moth-wise things on the morrow. We were not disappointed.

I suppose it is only natural to feel pleased with oneself when making a discovery. The steep side of Cranham Bank made the setting for this story, which concerns a classical piece of biological fact well known to every student from reading books or listening to lectures. I had, of course, no botanical instruction at school but during one of my first camping weekends at Cranham I caught some moths (I recollect the species as being the chalk carpet), and when I examined them something struck me as being odd around the mouth-parts. On examining the heads of the insects as minutely as I could, I made out a number of curious club-shaped growths clustered on to the base of the proboscis of each moth.

This phenomenon slipped out of my mind until later that day when I lay sprawled in the sun on the bank and gazing into an individual flower of a pyramid orchid, when to my astonishment I saw a pair of identical club-shaped organs to those I had seen on the moth tongues, situated on a little platform in the heart of the flower. It dawned on me that the mechanism was to ensure cross-pollination. To confirm it, I picked a piece of thin dried grass stalk and probed it into the throat of the flower, and both pollen masses (pollinia, I now know them to be called and as they proved to be), attached themselves to the grass-stalk when I pulled it out.

The rare and beautiful butterflies and moths and plants at Cranham and the surrounding countryside gave me immense enjoyment and I have been back a few times since, and I have been able to show my daughters the fly orchid growing in the identical situation on a very steep hillside where I had observed it thirty years before. But it has now been several years since I have visited Cranham. I could not stay away from Wyre Forest for so long.

When I wish to think of a beautiful and happy place then my mind turns to Wyre and to its running stream, Dowles.

Dowles Brook is formed by the union of three brooks at Furnace Mill. Mad brook comes in from due north arising from Bradley two miles away and rushing through Hall of Hammonds. The Baveney Brook comes in from the north-west running parallel to the river Rea that drains the east side of the Clee Hills. It rises at Bagginswood to the west of Kinlet Park, a long northwards-pointing spur of woods and parkland with its origin in the forest. From the south comes Lem Brook, starting as Tanner's Brook and rising from Clows Top. Water running down the north side of the village street finds its way into the Severn via Dowles, but on the south side it runs into the Severn via the river Teme. The Gladder and Dick brooks rise on the south side of the Bewdley to Clows Top road and drain the Callow Hill side of the forest. The word 'Gladder' lives on, although the yellow flag-flowered rush, the Gladdon, alas, is no longer to be found at its edges.

From Furnace Mill to the Severn, just about a mile above Bewdley, the Dowles runs its course of about four miles, if short cuts are taken down some of its twists and turns. Here are four miles of some of the loveliest forest-stream scenery that can be imagined.

The ancient name of *Dulas* meant dark, for dark and cold it is. For many miles it flows over a succession of pavement slabs of greyish-brown bunter sandstone, at least in the straight runs, but where Dowles twists, the sandy bank is cut away leaving the stalactite roots of heather, gorse and oak searching and thrusting for fresh support. But support they do not get in the winter floods. They fall across the stream, dragging lumps of heathery peat with them and tangle in their branches a vast agglomeration of straw-like creamy twigs. Even in the heat of summer, when the silver-washed fritillary butterflies are fighting for the thistle heads, the water of Dowles is ice-cold and a gasp cannot be suppressed when a first bare foothold is sought on the slabs.

The island, as we called it, was not a true island, but it gave the impression of one. The Dowles Brook ran almost in a full circle of about a hundred yards' diameter. A small fast tributary stream ran into the brook at one of the elbows, but at some time in the past it had run into the other elbow and sometimes used both channels during the winter floods. The most remarkable feature of the island was the trees. I have no idea how they came to be planted there, but I have never seen such a magnificent group of spruce, Scots pine and oak. My happiest boyhood days were spent there, either weekend camping or camping for longer holidays. For some years the island was the home of a pair of long-eared owls. In 1927 they made their nest or, more accurately, laid their eggs in the hollow between two roots of a giant spruce tree. It was dark under the tree and the needley earth was dry and dusty. The owl

The little yellow mouths of cowwheat (Melampyrum pratense) *always face one way and are found only where the oak leaves have formed a soft mould for its hair-like roots. It is said to be partially parasitic, its roots deriving nourishment from sucker-like protrusions which fasten on to the roots of grass, but I never found it so.*

The world suddenly seemed to enlarge when I caught my first pearl bordered fritillary butterfly, at the Lickey Hills within the city of Birmingham. Previously I was only familiar with the garden whites, the small tortoiseshell, peacock and red admiral, and the butterflies of the swamp at Great Barr, the green veined white, the small and the common blue; but here was a butterfly of the unpolluted woodland where the violets grew on which the caterpillars fed. I have somehow always associated the pearl bordered fritillary with the natural English woodland untouched by the dirty fingers of industrialism. It seems quite natural to me that I should have been searching out this speckled fulvous orange butterfly along the banks of Dowles when I met Emma Fischer, who asked whether she could see the butterflies that I had collected.

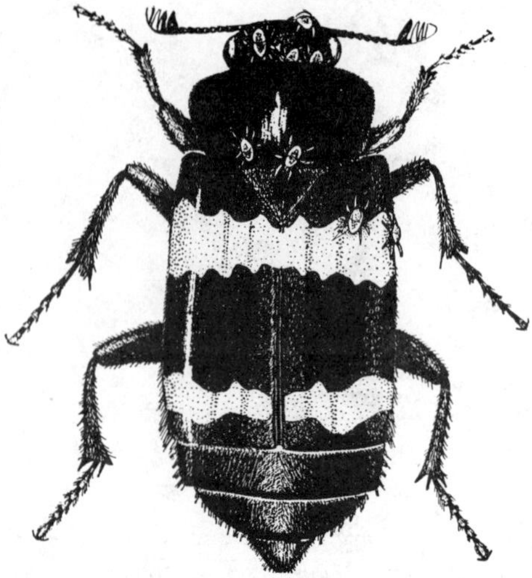

The black and orange banded burying beetles fly considerable distances in search of a site for their eggs. This site is the body of a small mammal such as a mouse or mole. The underlying soil is scooped out and the little body sinks into its grave then the Necrophorus *lays its eggs thus ensuring a meat supply for the developing larvae. I once found a* Necrophorus *investigating a broken egg which had fallen from a pigeon's nest. Of the three British species the one we find around the forest is called* investigator *and it often carries around with it a number of passengers, mites of several species, able to cling on to the beetle and thus widen their distribution. One* Necrophorus investgator *I found in the forest was carrying 37 specimens of* Poecilochirus necrophori *and 7* Macrocheles glaber.

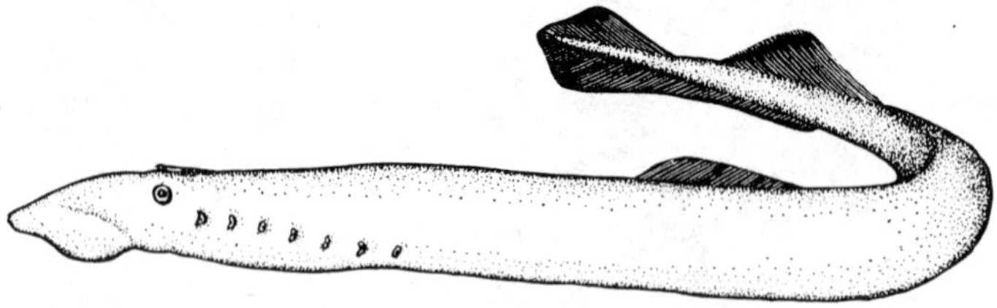

I first found the brook lamprey in a Wyre Forest streamlet when I was a boy, and since that time its biology has alternately enthralled and repelled me whenever I have come across it. Its bag-like sucker mouth and encircling teeth enable it not only to parasitise other fish—chewing the living flesh from them—but by adhering firmly to its host it gets transported by migrating fish. It is unique amongst fish, the bag-fishes excepted, in not possessing jaws, but its pineal organ or 'third eye' is distinctly visible on the top of the head. Although this is sensitive to light it is thought not to be an organ of sight. But the most remarkable thing about the lamprey is that it spends the first three or four years of its life as a peculiarly immature larva-like form inhabiting a tube buried in sand or mud and living on organic matter which it is able to strain out from it.

left her two white eggs with a frightening clap of her wings and her erect 'ears' were like the horns of the devil.

The Long-eared Owl was known to nest in the forest in the 1890's. This is a bird of the conifers and is seldom seen by day on account of its habit of roosting close to the trunk of a spruce or pine. It was reported in 1963 as nesting in the forest near old haunts that are today more thickly planted with conifers, but its eggs were taken.

Under the spruce the ground was quite bare and a lovely warm cinnamon brown, but around the oaks was a carpet of grass, primroses and cow-wheat. Cow-wheat is a plant always associated with the forest and whenever I see its delicate line growing in Ashdown Forest, Sussex, or in North Wales, I feel a little tug at my heart and I see the bare paths of Bunter sandstone slabs edged with this lovely plant in the semi-shade at the edge of the oaks, with its cotton-like roots only held by dead oak leaves. I can hear the bells on the forest-grazing cows. Where more light penetrated and in the grassy rides, the little purple blobs of self-heal came into their own from midsummer onwards, but before this, at the time when the two species of pearl-bordered fritillaries were flying, bugle was everywhere in drifts of fat grey-purple spikes.

If we except the *Fontinalis* moss which here and there abundantly encrusts the stones, there is no vegetation in Dowles. But where Dowles empties into the Severn the river is wide and shallow, scarcely a foot deep at any point, and there are long trailing manes of water crowfoot and in the early summer the whole surface of the Severn is whitened like snow with its hawthorn-like blossoms. But if Dowles contained no water plants there was certainly no lack of aquatic animals. Three or four brown and speckled trout could be seen in every little run of the brook and in the deeper pools were small shoals of silvery chub; or rather they could be seen if the viewer approached with extreme caution. Whether the members of the fishing club who rented the fishing, the White Swan Piscatorials, ever caught any exceeding about nine inches in length, I do not know, but that seemed to be their general maximum size. For some years a pair of herons fished for them also, and my heart stopped still as I came on them unexpectedly as they rose up with some commotion and awkwardness through the dense alder branches that fringed the brook. In their nervousness, or was it a defence mechanism, they released a white spray which often spattered the bushes around me. Later I found their dead bodies riddled with lead shot and carrion beetles, that scurried away as I moved over the grey-clad cadavers.

Trout and chub were not the only fish in the brook. Loach and miller's thumb were also there, and eels too; once I caught a lamprey, seven or eight inches long, in a mug I had. Its sucking bag-like mouth gave it a somewhat repulsive appearance, but it was most beautifully suffused with carmine under the belly at the head end.

The lamprey of the gourmet is the sea lamprey (*Petromyzon marinus*) which formerly came into some of our rivers to spawn. It is now a rare fish, although a few still come into the Severn. It grows up to three feet in length and is a parasite of fish to which it attaches itself by its sucker-like mouth. It possesses no jaws, but its sucker is lined with sharp teeth. The basking shark is often attacked by it. But in my search for caddis grubs in the streams of the Midlands I have on a number of occasions come across the brook lamprey (*Petromyzon planeri*). Those I have seen have not been longer than nine or ten inches. It occurs in the little streams that run into Dowles and I have accidentally caught it in my caddis net in a stream at Temple Balsall in Warwickshire, when pushing the net through a dense mass of aquatic plant life. Once I saw several hundred undulating their way against the current in a stream in Gloucestershire, on migration to spawning grounds. There is a third species, the river lamprey (*Petromyzon fluviatilis*), although some authorities consider all three species to be, in fact, one. The old Gloucestershire name for a lamprey was a 'pride', which gave rise to several family names, including that of my mother—'Pridey' or 'Priddey'.

The large flat slabs on the bed of the brook made ideal hiding places for many different sorts of animal. The miller's thumb often laid their eggs in convenient hollows under the stone, and when disturbed by the rude turning over of the stone they seldom swam away. The loach were not quite so easy to catch. They were mostly found where there were sandy spits. They just eluded the outstretched hand by swimming a few inches forward and seeming to disappear in a little flurry of sand.

Crayfish were very abundant and always seemed to be of two distinct sizes. The large black ones about four inches long made quite a commotion in the water as they flapped their way backwards from under one stone to another and they could give a finger a reasonable nip. The small ones were about an inch in length, olive green on their backs but with their claws dipped in orange red. They shared the same slit-like caverns under the sandstone slabs as the *Stenophylax* caddis and often a dozen or more of the latter were congregated together.

Those summery weekends in the forest were a great delight throughout my youth, and, as if by a miracle, I continue them now—as I shall tell later.

Before the giant trees were cut, our little bivouac tents were unstrapped from our backs and erected within two or three minutes, and our frying pan and kettle conveniently stuck among the twisting roots of an oak balancing on the very bank of the Dowles. The ground was hard but flat, and we cut bracken and silver-birch twigs on which we laid our cycle capes. Water we fetched from the tributary stream in an old canvas bucket and we made a fire. I cannot remember it ever raining and seldom were other human beings observed in the forest.

Sitting round the fire at night under the giant trees we watched the shadows on the oak-clad hill rising up to the railway line. The bracken clock purred away or made quick, flicking flights against the deep green sky. When the last sparks rose up we turned shivering to our home-made tent. As we lay on the ground, the noises of the forest were magnified a thousand times, the calling owls, the far-off farm dogs, the breaking twigs and sudden nearby scurrying of deer, and on Saturday nights the sound of a concertina from a forester's doorstep carrying down the valley and over a mile away.

When I first heard the nightjar churning on the darkening bracken-clad slopes behind the island I thought it was an insect. I vaguely remembered reading something about the 'bracken clock', and this seemed to fit the description, but when I had cycled home I went through my books and found indeed that it was a bird. During the next visit to the forest I saw it flying around the scrub of silver birch after the moths and coming to rest lengthways on the branch of an oak. Professor Steele Elliot estimated in 1930 that in Wyre Forest there was a pair of nightjars to every hundred acres. The forest was at that time much more open than is now the case and it must have been even more so at the end of the last century when the black grouse were resident in the forest. Only the bilberry and the heather on which they fed now remain.

One afternoon, four of us, returning from Sturt Common, were astonished and pained to see two girls sitting deep in the bracken on the bank of the Dowles within a few yards of our tents. We all thought this a piece of impertinence, especially as they did not move off when we went to our tents. But when we had made a cup of tea we walked over to them with a cup each and I apologized for the swirling specks on the surface with, 'Please excuse the flora and fauna.' A prophetic remark, as it turned out, as I eventually married one of the girls—Emma Fischer by name—and on many occasions since I have begged forgiveness for the presence of my Ptinid beetles, my caddis worms in dishes about our home and my chunks of woodwormy wood. At that first meeting no arrangements were made for another meeting and two years passed before we did meet again, when I was on my own in the forest in almost the selfsame place.

The small pearl-bordered fritillaries were flitting around the violet bugle spikes, always so luxuriant along the path and straying in little courting groups on to the crushed raspberries and cream of the ragged robin in the damper rides. The previous year a white variety, as well as a short series devoid of the spots on the top wing surfaces, had been taken and I was looking for one of these. As they came into my net I was busily examining them for lack of spots when I vaguely became aware that I was being watched. Down on the bank of the Dowles, with the steep oak-clad bank behind, and the young uncurling bracken around them, were three laughing girls. Bare-legged, with toes dangling in the brook, they turned towards me and some comment passed on my butterfly net and laughter followed, but one girl sprang to her feet and ran over to me and asked whether she could see my butterflies. It was Emma Fischer again, but this time I found out where she lived and we arranged to meet.

4

A Do-it-yourself Degree

I HAVE already made some comments on a few aspects of getting an external London University degree, but I thought it might be of interest to set down some details on how a large number of young men before the last war, but a much smaller number since, obtained an academic qualification without making a full-time attendance at a university (or, indeed, any time at all), with the exception of attendance necessary for examinations.

The first stage after matriculation was the London University Intermediate Science examination which required four subjects to be passed at one and the same time. In industrial laboratories it was the recognized thing that the subjects taken were chemistry, physics, pure maths and applied maths. Accordingly, and in conformity with convention, I applied myself to these subjects and commenced as an evening student at the Birmingham Central Technical College.

The 'Tech', as it was universally known by those who used it, was a triumph of men over a building. It would be hard to imagine an uglier edifice. From the railway, which seemed to bury itself in its very bowels, it was a great anonymous block and for all its warren of rooms it would be even harder to imagine a less pretentious entrance. The steep sloping street on which it was built meant that all the steps ran off unevenly. You went into the small stiff swinging doors off your balance and it certainly took a prodigious physical effort to get inside those doors. It always left me wondering whether the builder expected a siege, to prevent the students from coming in, or perhaps to prevent them ever getting out.

Evening class started at 7 p.m. and finished at 9.30 p.m. In the experimental sciences there was an hour's lecture followed by an hour and a half's practical work in the laboratories. It was most unusual, at intermediate level, for the classes to be anything else but full. The pure and applied maths classes overflowed. On several occasions at the beginning of term I had to join those standing near the door unable to find a seat. Practical notebooks were very important at the Tech, no loose-leaf books, scrappy bits of paper, nor indeed any other notebook but the standard college type for the particular science. They were very large, bound in thick card in a deep maroon colour with the college arms embossed on the front in gold. For the biological sciences they consisted entirely of good thick cartridge paper, but for physics the cartridge paper was interleaved with ruled paper for descriptive notes. The strictest discipline was exercised in practical work. The evening experiment or dissection had to be finished and written-up that same evening. It was not much use saying that you would finish the labelling at home.

There was one very important perquisite allowed to us by the company when we had to take our university examinations in London. This was to transport us to London by car to the company's depot at Albany Street. In order to test tyres, a fleet of motor-cars was run around the clock, in three eight-hour shifts and operated by a very competent team of 'test-drivers' as we knew them. Some of them were ex-racing drivers and it was always a joy to be driven by them—so confident, skilful and safe were they. On one occasion the engine of the car I was travelling in, up to London, failed, but my driver waved down a long-distance bus going our way and I got to London by that means.

When in London for exams I and a number of others from the lab stayed at a guest house at Stockwell, near to Brixton, which was very convenient for the underground. Many examinations were held at the Imperial College of Science in South Kensington. To save buying meals in London we often took a large parcel of sandwiches. On one occasion I remember I was in the Science Museum surreptitiously eating lunch from my pockets with three other Dunlopians, after a morning's chemistry session. The new public-address system 'wireless' had recently been installed there and 'Fingal's Cave' was being broadcast. Suddenly, apropos of nothing, one of my companions said 'I think I did that question very well.' 'Which?' we asked, and he replied, 'The one on Mendelssohn's periodic system, of course.' 'But it was Mendeleef, not Mendelssohn,' we all chanted. 'Good God!' he said, 'I put Mendelssohn.'

I could just about pass in chemistry and physics, although in the four times I took London Inter I did not always do that. But pure and applied maths almost always beat me, although once I passed in pure maths. I could learn the formulae, but I had the greatest difficulty in applying these to actual problems. When at last I passed London Inter it was after I had made the decision to take a biology degree and I passed Inter in botany and zoology with physics and chemistry. Incidentally, the six-guinea entry fee took some finding each time and I usually had to sell something to make up the money. My more valuable possessions went one by one to raise the wind, a collection of postage stamps at one time and a sort of cannibalism at another because I sold a slide rule and a special *verplatiniert* box of chemical weights. I gave up a lot of tangible for the intangible.

I was twenty-two when, as I have already intimated, I made the decision to go all out for a degree in biology and to have another go at Inter, but this time dropping the mathematical subjects and substituting botany and zoology for them. It was during the period when I was trying to make this decision that I had met Emma and we were meeting regularly. Then, in the September, on enrolment evening I met Ernest Moore, the senior lecturer in the department of biology at the Tech. I unfolded to him my plans: how I wanted to become a professional entomologist with a university degree. I must have told my story quite well, and with the heads of departments sitting uneasily in schoolboys' desks I was duly 'enrolled'.

Although I had read very widely in natural history I was unfamiliar with the routine and conventional knowledge required for botany and zoology, but I was a very willing pupil. I very soon found out how useful was some ability to draw quickly and accurately, and I made the best use of my early training and my personal inclinations in this respect. Besides Ernest Moore there was Pat his wife, who played a large part in the evening botany teaching, and there was Cosway, who did the greater part of the zoology. The Moores had a remarkable effect on my career as a biologist—I dreamt such dreams but they made them come true.

After a few months Ernest Moore took me on one side and asked me how important it was that I got an external London degree. I told him that it was very important and it was really the thing I wanted to do. He then said he had been discussing me with his wife and they had come to the conclusion that if I was prepared to do what they asked of me in the way of work they thought that I could accomplish it. I found

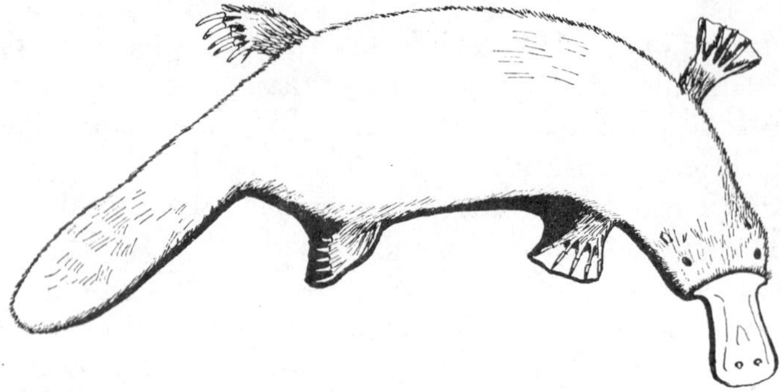

As soon as I could read I started to borrow books on Natural History from a branch of the City of Birmingham Library at Perry Barr. The works of the Rev. J. G. Wood, full of very detailed steel engravings, were very familiar to me as were the works of many other similar authors, but when subjecting myself to the discipline of reading for an external degree, involving the subject of Zoology, this meant making a series of drawings of the special features of the world's animals. This was mostly carried out at the Natural History Museum in Birmingham where this duck-billed platypus resided in a glass case.

A small difficulty associated with drawing animals in glass cases was the restriction of movement. One could not open the case and move the specimen to a more convenient position so that all the features it was desired to draw could be seen. Nevertheless I managed to get some specimens arranged so that I could effect a conventional drawing.

I have never visited Australia where the marsupial fauna is of such interest to the zoological student. My spiny anteater of Australia was, alas, drawn from a stuffed skin but it played its part in filling me with wonder at the evolution of animal life.

this very encouraging and a source of strength, when, night after night in all weather, I made the cycle journey up to town. I enjoyed the classes immensely and I did my best to get all my homework done, and although for most sessions I went four nights weekly, for one session even going five times each week, my attendance record was good. One piece of advice that Ernest Moore gave me concerned, in his words, 'extraneous activities'. I must drop them. This involved membership of the Birmingham Natural History Society of which I was an active member and a number of other clubs and societies to which an active young man was likely to belong. Rugby was the only activity which I kept. I rather think Pat Moore had caught sight of me talking outside the Tech to the same girl on several evenings and had wondered whether this siren was going to seduce me away from my work. In fact the seduction was quite painless. During the week we would meet only outside the Tech when Emma happened to be at her psychology or cookery class. Often we would consume an evening's work and pass judgement on it before getting on to our bicycles and going our separate ways. This continued for four years, but some Sundays, if not too near examinations, we would be off to the forest.

After getting Inter out of the way I was all set for the real task. The particular degree of London University that I decided to go for seemed very appropriate for the sort of study I wished to make, and, furthermore, fitted in perfectly with my own natural inclinations. This was called the 'special' degree in which I was first to take a subsidiary final subject, in this case botany, and when this had been passed I was to take a final subject, zoology, together with a special subject. One of the listed special subjects was entomology. I then registered for this sequence of examinations, the finals in zoology and entomology to be taken in 1936.

There were two other students for subsidiary botany and together with two other students doing microscopy we were able to attend the Technical College for practical work, using the back two benches of the biology laboratory when other classes were being held. I have no doubt now that this arrangement was highly irregular, not only did we not constitute a minimum class of eight but we were probably tagged on to another class. I am very sure today that the more rigid rules would have precluded our final botany class being held.

Ernest Moore would make a selection of microslides he wanted us to examine, study and draw in our practical notebooks and occasionally

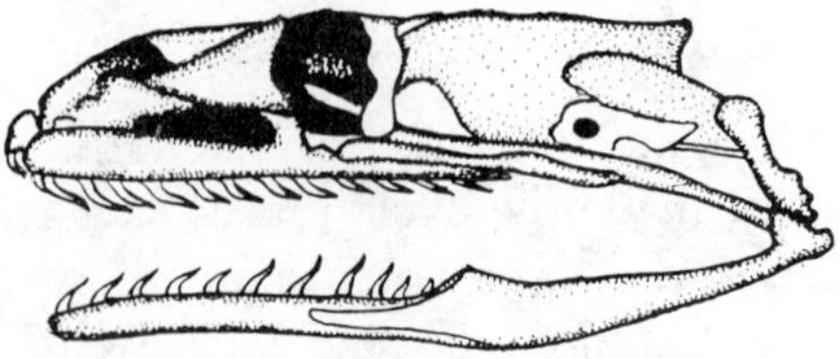

Snakes have always fascinated me. When I saw my first adder curled in the sun near a pile of dried-out brash washed down Dowles by flood, we watched each other for a long time. Later I had to study the comparative anatomy of the snake and I drew the skull of this python. I find that contemplation of the evolution of the snake, so perfectly adapted as a swallower and digester, chills my spine.

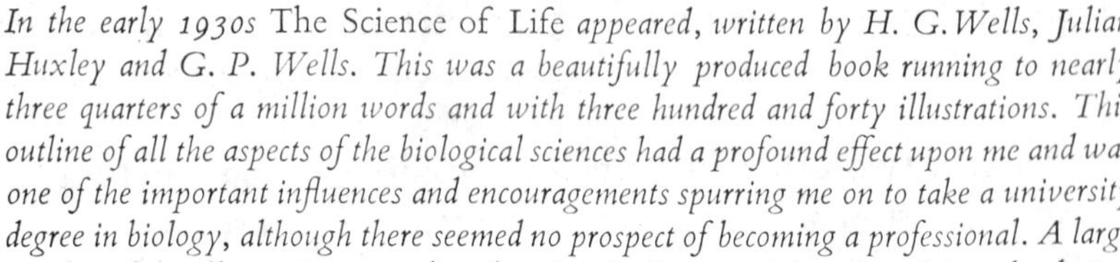

In the early 1930s The Science of Life *appeared, written by H. G. Wells, Julian Huxley and G. P. Wells. This was a beautifully produced book running to nearly three quarters of a million words and with three hundred and forty illustrations. This outline of all the aspects of the biological sciences had a profound effect upon me and was one of the important influences and encouragements spurring me on to take a university degree in biology, although there seemed no prospect of becoming a professional. A large number of the illustrations were line drawings by L. R. Brightwell and I much admired them.*

Unconsciously or consciously I modelled my own drawing style on his and this is very evident from my notebooks kept when I took the course in marine biology at Plymouth. The great wealth of marine life was quite new to me and I stayed up very late each night drawing animals I had never seen before. This drawing of the cuttle fish Loligo media *is an example of Brightwell's influence on me.*

an opportunity would arise for a short lecture and we gradually made our way through *Strassburger's Textbook of Botany*, a copy of the 1930 edition of which we each possessed. Until that time this book was the most comprehensive and detailed book I had ever bought (at thirty-six shillings for 816 pages with 861 illustrations) and has been a valued friend ever since. Ernest Moore did not consider that he could cover the syllabus for the examination and advised us to take the correspondence course in final botany of the University Correspondence College. This involved a strict discipline in written work. The list of reading, together with question and essay subjects, would arrive one week and had to be posted back the next. The corrected answers would be returned, and mine usually had plenty of red ink on them, which depressed me very much, because when I wrote the answers I invariably had the textbooks open before me. I naturally wondered what was going to happen at the examination when I did not have the benefit of the textbooks. However, I persevered. About a month before the exam Pat Moore gave me a list of twelve essays, saying that she had been through the examination questions of the last ten years and that she was sure that about fifty per cent of the exam would involve the subject-matter of the essay questions she had written down. Accordingly, I spent the last month answering Pat Moore's hypothetical examination questions as revision. Her prophecies turned out accurately and when I had written the last word of the examination answers I was confident of passing, which proved well founded.

The final zoology class was held also at the back of another class and there were only three of us. The lecturer took us through a large number of dissections of animals in the London syllabus, but not quite all. In the reptilia, although I dissected a lizard and a snake, I did not cut up a tortoise and by a stroke of ill-luck for the practical examination I had the latter animal given to me to dissect, in order to show the principal blood vessels. This came as rather a blow, in fact often when confronted with the way in which the quirks of fate twist our tails I look back to my own confidence as I went into the examination room and to the way in which it immediately dispersed when I saw the tortoises on the benches. To me now the tortoise is the symbol of deflation. Perhaps, as in the emblem of the Fabian Society, the tortoise has the motto, 'When I strike, I strike hard'. This animal certainly struck me hard.

A series of microscope slides and specimens of the invertebrate animals were obtained for us, and in addition permission for us to visit the Zoological Museum of Birmingham University on three Saturday afternoons. Sunday afternoons I spent at the City Natural History Museum, drawing the skulls and other skeletal features of the vertebrate animals, and I accumulated a good collection of labelled sketches of the reptiles and mammals and the various adaptive features shown by birds. In addition to this a University Correspondence College Course was again taken. The written work was very cleverly worked out by the U.C.C. tutors in that the answers to their questions could not be found in the standard textbook, the two volumes of Parker and Haswell. This also had a depressing influence on me, as I thought that my reading was not of high enough standard, and every week I had to spend more time at the City Reference Library looking up papers in journals which, at that time, I thought obscure. In this subject too the marking of questions was severe and obviously carried out with great care. In my answers I made use of sketches and diagrams as often as I could and generally got to the required standard of the course, although again I usually had the opened textbooks and papers in front of me when I was answering the correspondence course test papers.

It was most fortunate that I was able to gain a place in the Easter Vacation Course run by the Marine Biological Association at their

If the recording of biological phenomena is to be of any significance it must be accompanied by essential information. Thus the reporting of a specimen must be associated with its exact position, including its parish and county, the date and the name of the captor. Abbreviations are dangerous, being familiar only to the original collector.

During my life, tens of thousands of specimens have been sent to me for identification, mostly wood-boring insects, and occasionally very rare insects have been sent which have caused much time-wasting annoyance, whilst the essential information has been extracted from the sender. Very often specimens are sent through the post in an ordinary envelope and arrive in the condition known as H.O.F. (hammered out flat)! My drawing of the sea slug Aeolodoa papillosa *(centre) is in honour of the Plymouth laboratory of the Marine Biological Association who first taught me these things.*

Plymouth laboratory, the Easter before I took the final examination for the Bachelor of Science degree of London. This course was run by two marine biologists on the staff of the laboratory, Douglas Wilson and G. A. Stevens. The first named was a well-known photographer of marine life, whilst the latter was a practical fisherman, and both were well versed in the biology of life in the sea. All the course students were from universities except myself, and I felt a little inferior because of this; they all seemed so full of confidence and so clever at repartee.

The course was conducted in a wooden hut at the back of the aquarium on the Plymouth Hoe. It was the first course of its kind that I had ever attended and the lessons of organization learnt were to play an important part in the courses organized for the new technology of wood conservation in later years. We went out and collected marine animals from the rocks, the rock pools, under stones at low tide, trawled in the trawler *Salpa* in several fathoms off the Mewstone Rocks or in the deep pools for crinoids. We brought our captures home in jars and buckets covered with seaweed, to identify them with the keys from the standard works. I made a resolve to write up a good notebook and to draw the maximum amount. We had a good grounding from our leaders in many features of marine life and it all helped a great deal when it came to my final zoology examination a few weeks later. One bit of good fortune I well remember was that at Plymouth I collected a parasite, the leech *Pontobdella*, from a skate that we had trawled up in the English Channel and it turned up for identification in the examination room. Neither of the other two zoology aspirants were able to identify it, although this did not seem to make much difference to their subsequent careers, as they both became professors, one in parasitology.

I cannot recall who it was who suggested the lines on which my study of 'special' entomology should be pursued. I rather think it was the U.C.C. to whom I wrote enquiring whether they ran such a course. However, I am very grateful to whosoever it was, because I enjoyed the study so much. Two principal books were suggested: *The General Textbook of Entomology* by A. D. Imms was the basis of my reading. I had the third (1934) edition and this book has never been far from my right hand ever since, except that the latest edition (now the ninth) has been substituted as it appeared. This great book is a masterpiece of condensation and abridgement. It is said that the number of known insect species is in the region of a million and the complexities of anatomy, physiology, development and classification of this great horde of living things is dealt with in 886 pages and 606 illustrations. It was a source of great encouragement to me when I first saw my own research work on the caddis-fly biology referred to in it.

The other principal book was *The Textbook of Practical Entomology* by Professor Frank Balfour-Browne. This work is in three parts, the first part of which commences with the general structure of the cockroach, the second part an advanced course which commenced with a very detailed dissection of the water beetle *Dytiscus marginalis*, whilst the third part exemplified the principles of systematic entomology. This last part was a course for graduates who had applied for colonial posts in economic entomology produced by Professor Balfour-Browne when he went as professor to the Imperial College of Science from Cambridge (where he was a lecturer). This book outlines and gives useful information, practical laboratory-wise, on about 600 dissections, whole and part insects, and during the two years in which I practically lived with this book I made 466 labelled drawings of the required observations and dissections.

All the stipulated specimens I collected myself, although I was able to get a few slides from commercial sources. This meant looking up the insect in the books and finding something about its biology and distribution. It was a formidable task, but on the other hand was a very worthwhile exercise because it gave me a very wide spectrum of information about insects that was bound to come in useful to me if not for the examination itself. In the case of some insect specimens that it was quite impossible for me to collect I wrote to influential entomologists whom I thought might be able to help me. I thus came across the freemasonry of entomologists anxious to help with specimens, which delighted and encouraged me and since that time I have always endeavoured to follow their example. Now that certain specimens inevitably come my way, I am myself often in a position to help someone else. Having to collect all the specimens required for working through Balfour-Browne's textbooks was, of course, a time-consuming operation. When the old ache came along, however, compelling me to walk along Dowles through the forest, I felt I could let go, so long as I had a reasonable chance of achieving the objective in the form of insects required for study. Emma, of course, came with me.

As well as the time taken in collecting the specimens, further time was required in making the necessary microscope mounts from them. This was, of course, so different from the university departmental practice where all the mounted specimens available are nicely labelled and indexed.

In the end I got a pass degree in the 'special' examination and my salary was raised at Dunlop's from £3 4*s*. 6*d*. to £3 7*s*. 6*d*. per week. Emma and I were married in the October of that year, 1936.

Because examinations, both university and tech, were mainly in July it meant that for some years I lost touch with the forest in June and July. It was in spring and autumn and even in winter that I maintained my contact during these study years.

Surely there is no more lovely sight than a forest meadow in early May covered with early purples and cowslips. The deep maroon of these orchids show up in the pure saffron of the other flowers and their lime-green stalks. I first met the early purple in a meadow near the Dowles Brook where it runs into the Severn and when the orchids are in bloom, pairs of yellow wagtails having newly arrived, fly skittishly along the brook and seem to be excited as they re-explore every little stretch of the stream, performing little aerial feats in endeavours to catch the early-flighting stoneflies.

The dipper is also prospecting and in mid-April is flushed from under the overhanging banks, but although a much heavier bird than the wagtail, its flight is unobtrusive, swift and purposeful, with no hint

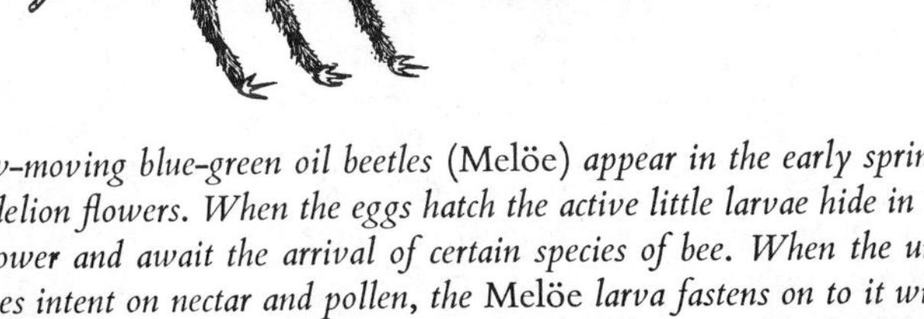

The slow-moving blue-green oil beetles (Melöe) *appear in the early spring with the first dandelion flowers. When the eggs hatch the active little larvae hide in the recesses of the flower and await the arrival of certain species of bee. When the unsuspecting bee arrives intent on nectar and pollen, the* Melöe *larva fastens on to it with its three pronged claws (such a larva is called a triungulin) and is carried back to the bee's nest. Here it changes its form, becoming fat and sluggish as it preys on the bee's larvae. I found oil beetles in Sutton Park, but I did not learn of their biology until I had to draw a triungulin when I was working through Balfour-Browne's textbook of Practical Entomology.*

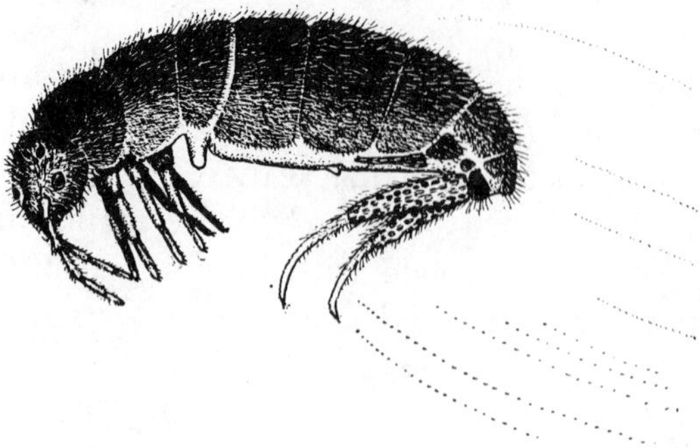

The debris of the forest floor is abounding with life. Dead leaves, twigs, branches and trunks provide sustenance to a host of living organisms which, in turn, transform the products of the forest's death into the lifegiving nutrients of the soil. The agents of wood decay, the wood-rotting fungi and the wood-boring insects, thus provide space for new life, enabling new trees to grow in the place of those that, having passed maturity, exhibit senescence. Amongst the damp litter and crumbling earthy wood-mould live the springtails. They share with the bristletails, like Dilta *(see page 83, top), the privilege of being amongst the earliest of insect-like forms still living. They were the forerunners of the teeming insect species that now inhabit the earth. But, although the little springtail has not evolved very far along the line that leads to the possession of wings, they are present in incredible numbers on the forest floor.*

The Mallophaga *is the name given to the 'biting' lice or 'bird' lice. About 1,700 different species of these very small insects are known and although the great majority of them are parasitic on birds, a few species are pests of mammals. Strangely, all domestic animals have 'bird louse' parasites. Although occasionally they will imbibe a little blood, if they get the chance (and this is usually only by accident), their normal diet consists of bits of dead and dying skin and bits of feather or hair. They are of interest to the ornithologist as pointing to the evolutionary relationship between birds. Although one cannot help showing a degree of revulsion to these creatures, their adaption to their environment is incredible. See how the goat louse can cling to a single hair of its host by its interlocking mandibles, its six legs and even the terminal segment of its abdomen.*

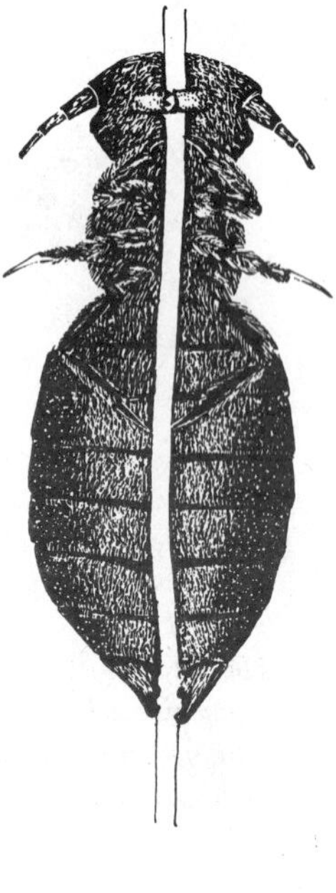

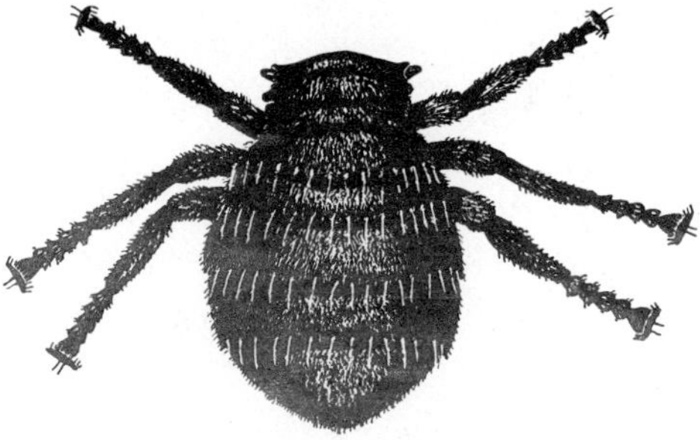

The little bee louse, Braula coeca, *is a dipterous fly but it possesses neither wings nor halteres. Its comb-like tarsal claws enable it to hold on to the hairy thorax of the honey-bee where, in the tight fit of a flower's corolla, it could be jostled off on to another bee and thus effect its distribution into another hive. My drawing of this insect and its inclusion in my collection was made possible by the generosity of Guy Morrison of Aberdeen, and is an example of the generous feeling which exists between entomologists.*

of the ostentatious enjoyment shown by the wagtail. A long still wait on a high bridge may give a glimpse of the dipper running over the stream bottom catching caddis worms by twisting the stones over, as I once saw from Ludford Bridge at Ludlow.

Earlier still, as soon as the first flowers open in spring, the 'bee-flies' appear. The large bee-fly *Bombylius major* is common round the forest from the first sunny days in March to mid-May. Its agility on the wing is unsurpassed amongst the dipterous flies. It suddenly appears at a flower, a primrose on a sunny bank or at the aubretia in a forester's garden. It hovers with such precision that the wings are invisible, yet every hair on its furry bee-like body can be seen. When poised for a second or so in front of the flowers of its choice it hovers near enough to the flowers to allow its front legs to grasp the outer edge of the petal, then its long dagger-like proboscis reaches down the calyx in a few jerky movements. Its wings never stop their vibration during nectar sipping and if it is disturbed its disappearing flight can scarcely be followed.

The October days in the forest were full of exhilaration, although the keen winds did not detach the washed brown leaves from the oak. They would remain another month before they formed little drifts among the Bunter slabs. Now, the scabious reigned supreme because most other plants showed only their ripened pods and fruits. They lined all the paths through the forest, and on the swaying blue-mauve blobs the comma butterflies clustered. The comma was not always so widely distributed throughout England as is the case today, but when it was a rare insect elsewhere it was always to be found in Wyre.

Both the roe and the fallow deer were present in the forest and were quite a common sight. The smaller roe were usually in threes or fours, but the fallow were in larger herds. Sixteen once picked their way past me as I sat under a yew tree. Early one crisp frosty Sunday morning, walking along by the railway-line with Emma—we had come to Bewdley by the early-morning fisherman's train—I remarked on the likeness of an old tree-trunk lying on a bank about fifty yards away to a fallow deer when suddenly it got up and shot away: it *was* a fallow deer.

Late October saw the fallow bucks at their best. Their horns were wide and full, their spotting clear and they were not always in such a hurry to get out of sight. The slots of deer, both roe and fallow, and no doubt other feral species now found in the quiet woods and forest of the Midlands, were to be seen on every patch of damp sand or mud.

The fisherman's train was a remarkably convenient method of getting a day in the forest. This was a stopping train from Birmingham to Bewdley that ran on Sundays. Leaving Birmingham before seven, it arrived at Bewdley about eight-thirty with a complete load of Black Country fishermen, complete with creels and rods. It returned about five-thirty in the winter and seven-thirty in the summer. This made possible a good walk in the forest at any time of the year and the return fare was unbelievably low even for the 1930's.

5

A Home Higher Degree

I HAD already expended my annual holiday in 1936 on the Plymouth course, but I managed to get a few days off for our honeymoon in October, which we spent at Cranham. After the wedding reception we took a train to Cheltenham, then we went by bus to Birdlip and walked the last few miles to Cranham village through the beechy woods with our rucksacks on our backs. We roamed about Painswick, Haresfield and Miserdon and the surrounding countryside. It was mostly cold and frosty, so that it would have been inconvenient to emulate James Elroy Flecker—'Have I not sat on Painswick Hill with a nymph upon my knees, and she as rosy as the dawn and naked as the breeze!'

We were very happy, although in 1936 we were worried by our thoughts on what the future held for us. But a stroke of luck had come our way. I accepted the offer of a part-time job at the Tech which required me to teach first-year botany and zoology two evenings a week for £1 15*s*. per evening. I had, of course, not done any previous teaching, but I was fairly fresh from the course myself and I had all my notebooks. I modelled myself on the lecturing style of Pat Moore. I think that she was the most accomplished and competent lecturer with whom I have ever come into contact. Another source of encouragement, although no monetary consideration was involved, occurred when Ernest Moore asked me to provide some zoological illustrations for his forthcoming book *Biology for Students of Pharmacy*. These were my first published drawings.

Although we had put down a deposit of £25 for a house at Hodgehill Common, not very far from Dunlop's, about a year before, it was not ready for us. The purchase price of the house was £450 and in this era of real speculative buildings it was not surprising that two builders had already been bankrupted on this estate. We were kindly offered the use of a house near the Birchfield Road by my Aunt Nance.

We accepted and our first six months of married life were spent there, which I celebrated by contracting pneumonia.

In the competitive world I was convinced that if I was to become a professional entomologist I must obtain the highest academic qualification of which I was capable. I was determined to do this and studied the University of London's booklets on higher degrees. Because I had taken the 'special' degree, even though I had only attained a pass, I was eligible to work for the Ph.D. degree without necessarily taking the master's degree first.

The University of London had a scheme to help external graduates to work for a higher degree, whereby an 'adviser' was appointed, and after the usual registration with the university I put in an application to take part in this scheme. One of the grave disadvantages of the external university examination was the sense of being completely out of touch with the university, but with the higher degree that was to some extent vitiated by this advisory scheme.

Eventually, Dr. O. W. Richards (now Professor O. W. Richards, F.R.S.), the reader in entomology, was appointed as my adviser and as I had become extremely interested in the biology of caddis flies (which I deal with in a later chapter), I wrote to him outlining my plans to work on the biology of their immature stages. Dr. Richards, however, did not think this was a good plan for a Ph.D. project and wanted me to work on the PTINIDAE. I took the reason for this as something to do with the name of the department of the Imperial College of Science, which is 'Zoology and Applied Entomology'. I was rather disappointed that I was not to work on the caddis flies, as I had really become engrossed in them, but I was very pleased that Dr. Richards was to help me get my higher degree because I knew he was outstanding in the entomological world and enjoyed an international reputation. I went to see him at his room at Imperial College and found him to be tall and grave with a mass of prematurely grey hair. He was rather shy, but there was no nonsense about him, no time was spent on trivia, and he quickly sketched out a plan of work which he thought, if I accomplished it satisfactorily and if I wrote it up in a straightforward workmanlike style, would earn for me my higher degree. He gave me some leads on carrying out a literature search and where I could obtain living specimens of the beetles from which I could breed cultures.

The PTINIDAE, or spider beetles, is a large family, well over 400 different species being known throughout the world and placed in about sixty genera. In the present century several species have increased their range of distribution through commerce and can now be looked upon as cosmopolitan. The economic importance of the spider beetles lies in the fact that many species are pests of habitation, infest stored food products and do considerable damage to a wide range of vegetable and animal products associated with man, e.g. drugs and textiles.

In Britain fourteen species are found which are placed in seven genera. Some of these are very common, often being in our homes in extraordinary numbers. The spider beetles are small, rarely being as long as a quarter of an inch, and are usually somewhat globular in shape. They are related to the ANOBIIDAE (the furniture beetles) and share with them the character of the pronotum, the first part of the thorax, having the appearance of a cowl extending over the head, so that only the thread-like antennae of the head are visible if the insect is inspected from the top. From the biological point of view the spider beetles had already attracted some attention due to their having been reared through one or more generations in materials of seemingly low nutritive value and in receptacles allowing only very little gaseous exchange for respiration.

The army biscuits of the First World War contained 'maggots' which were the larvae of the brown or Australian spider beetle *Ptinus tectus*. The large-scale presence of the 'maggots' was notwithstanding the fact that the tins of biscuits were soldered down and thought to have been hermetically sealed. But what was found to be the case was that the biscuits slid down a chute after baking and thereby collected eggs of the spider beetle which had been laid in the farinaceous detritus in the sides of the chute and when the tin lid was soldered on there was ample air within the tin for the needs of the developing larvae.

The main claim to fame of another species, *Gibbium psylloides*, was that it was found in the tomb of Tut Ankhamen where it had been infesting the corn placed there 3,000 years ago in order to facilitate the celestial passage to paradise.

The eggs of the spider beetles are sticky and particles of foodstuff and miscellaneous debris adhere to them. Relatively small numbers of eggs are laid by any one female. The larvae are white fleshy grubs with the hinder part of the abdomen slightly enlarged and curled underneath it when disturbed, or when in the resting stage, and just previous to changing into a pupa. The larvae construct silken cocoons within which the change from larva to pupa takes place. The latter is white until

The galleries cut by the larvae of the beetle Melasis buprestoides *into fallen logs of oak and birch are characteristic not only to human beings but to the spotted woodpecker who tears off the bark and the pockets of rotted fungally decayed wood to get at the peculiarly shaped tadpole-like larvae. The galleries are not tunnels but slits, always transverse across the grain and the exit holes of the adult beetles are about one eighth of an inch in diameter about the same as the death watch beetle.*

I once picked up a small branch of silver birch lying on the ground on the bank of Dowles, it showed signs of attack by Melasis *and I took it home to see what would come out. As well as some* Melasis *beetles (as shown in the illustration) some alder wood wasps* (Xyphidria camelus) *emerged with the parasites* Aulacus striatus, *the first Worcestershire specimens of such species. But more remarkable still, four males and one female of an extremely rare sawfly* Pseudohemitaxonus sharpi *came out of the wood also. Before this, only eight female specimens were known, six British and two Latvian. No one has found the larvae but they are believed to feed on ferns.*

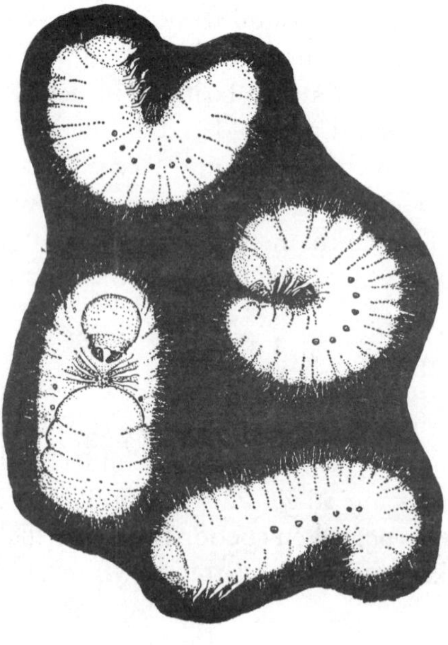

The small creamish-coloured grubs of the Spider beetles roll up into a ball when disturbed and their small weak legs cannot be of much service to them. In nature the grubs of the species Ptinus tectus *infest the small stores of grain and seeds made by the rodent-like marsupial animals in Australia. But in man's economy they infest a wide variety of stored grains and seeds and their products in all parts of the world where they have been carried by shipping. From 1936 to 1940 I spent a large part of my time, evenings and weekends, rearing large numbers of them in my home under controlled conditions of temperature, humidity and foodstuff, and my thesis for a Doctor of Philosophy degree of the University of London was based on this experimental work.*

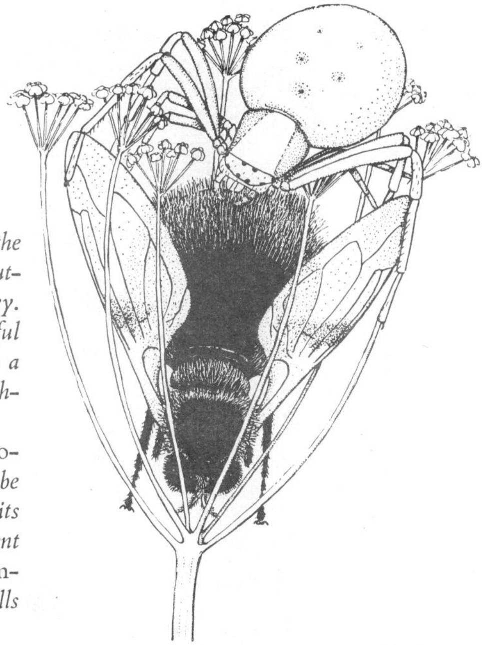

Armed with unsuspecting colour the Misumena *spider pounces with legs outstretched and sinks its fangs into its prey. The venom which is injected is powerful enough to quieten a large insect within a few seconds. This, of course, it does without the aid of an entangling web.*

When captured on the earthnut (Conopodium majus), *a large insect can be dropped into the cage of stalks and its juices extracted from this convenient position. The large fly,* Volucella bombylans L., *so like a bee, often falls victim to this species.*

shortly before the final metamorphosis takes place when it changes to a brownish colour. If the cocoon is disturbed, the pupa within it makes a series of violent flexions of the abdomen. After the splitting of the pupal cuticle the adult beetle remains within the cocoon for a few days before it bites its way out, during which time the parts harden. One, two or more generations may be found throughout the year.

The title of my research project was *A Study of the Food and Water Requirements of Ptinid Beetles*, and I was to breed some of the more important economic species of spider beetles in various foodstuffs and in a range of temperature and relative humidity. I searched through *The Review of Allied Entomology* volume by volume for references to previous experimental work of this type and to the spider beetles. In my thesis I see that I finally listed sixty-six papers and books in a bibliography of works that gave me information and help. I was able to get cultures going of four species, two of which, *Ptinus tectus* and *Gibbium psylloides*, I obtained from the field station at Slough of the Imperial College of Science. Mr. G. H. Mansbridge of I.C.I. supplied me with another species (*Ptinus fur*), whereas the fourth species culture, *Niptus hololeucus*, I got from specimens I collected from various

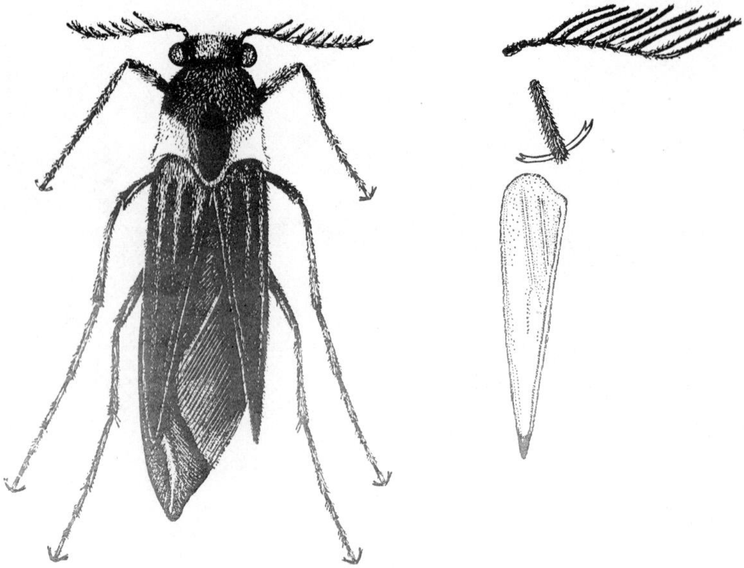

The recognition of a particular species of insect is a strange phenomenon exhibited by entomologists. Insect life is so rich in species that almost all entomologists specialize and take only one Order such as the beetles or the moths, of which to make a study. A number of professional entomologists study one Order during their working hours and another in their leisure time. However, when one stands on a flowery bank in high summer and the air is filled with such myriads of flying insects that there is a hum and whirr from their wings, one recognizes here and there insects with whose name and biology one is familiar. I recognize the beetle, Metoecus paradoxus, *the wasp parasite, when it flies along the hedges. When we dig out a wasp's next from the bank in midsummer we usually see a few pupae or adults of* paradoxus *ready to emerge from cells in the comb. Their feathery antennae, strange tarsal claws and sharply-pointed leathery elytra are so unwasp-like.*

situations. I used wholemeal flour in honey jars and kilner jars to get my stock cultures going and access to free water was given by keeping a small piece of wet cotton wool in a glass tube. Whilst my stock cultures were developing I paid attention to the means of rearing under constant temperature and various, but constant, relative humidities.

It was not until the early spring of 1937 that we eventually moved into our own new house at Hodgehill Common. It had two bedrooms upstairs and a small boxroom, and the latter was to be my 'laboratory'. I furnished it with a chair and a table on which sat my microscope and books and with all the bits and pieces I was accumulating that I would require for the experimental work. I made a few enquiries at the

Birmingham Central Technical College where I was lecturing as to whether I could use two constant-temperature ovens in the biology department, but my suggestion was turned down, so there was nothing for it but to construct them myself. My father made two boxes which fitted into each other and the intervening space was filled with insulating material. Underneath a perforated shelf was fitted a row of electric bulbs which were to act as the heating element. I was grateful to get a grant of £8 from the Dixon Research Fund of the University. This was my total indebtedness to public funds of any sort for my Ph.D. I also fitted a fan driven by a small motor which stirred the air and prevented hot spots. I constructed the thermostat myself from a design made by one of my colleagues. A large mercury reservoir was placed in the oven and a tube led to the outside, where a length of platinum wire was capable of being screwed down to make contact with the surface of the mercury. Another wire was fused into the glass tube at a level where it would always be in contact with the mercury. A relay switch of the rocking mercury type was connected into the circuit which shut the current off when the mercury level made contact with the platinum wire, and switched it on again when the mercury fell, losing contact with the platinum wire. This was a very simple arrangement and by present standards would be considered very primitive, but it worked especially well and the variation in temperature was well within the limits required.

Relative humidity was controlled by caustic-potash solutions of different strengths. A number of different salts and acids had also been used previously for this purpose, but caustic potash had the dual advantage that it does not give off noxious or toxic fumes when particles of food fall into it, and in addition absorbs the carbon dioxide produced by respiration. I had a little difficulty at first in getting the solutions of accurate strength due to the rapid carbonation of this material in air. After some experimenting I found that I could keep the relative humidities to within 1 per cent if the solutions were freshly made up every twelve weeks. This was with the exception of the 90 per cent relative humidity. Obviously this was because moulds grew vigorously on the foodstuffs under these conditions and the carbon dioxide produced by respiration caused rapid carbonation of the solution.

I used large sweet bottles with bakelite screw lids and a litre of the caustic-potash solution of the predetermined strength was used in each. The culture tubes were placed in the jars on a shelf above the solution.

These culture tubes were made from two-inch-diameter glass tubing which I cut from long lengths during one weekend. This was done by making a small file mark where the cut was required. Then strips of wet blotting paper a quarter of an inch in width were wrapped round the tube about an eighth of an inch apart. A blow-lamp was then directed on to the file mark and like magic a crack quickly encircled the tube. I got great satisfaction from this. I made brass rings for holding bolting silk at the top and bottom and I was then ready for the experimental work. One setback I had was to find that spider-beetle larvae could bite their way out of the bolting silk and I had to use thin lead foil on the bottom. I would mention that in order to get standardized results as far as possible and repeatable if necessary I had looked into the question of a complete description of the foodstuffs that I was going to infest with the spider beetles, and I had to make a simple study of the flour industry. I used plain flour (containing proportions of Canadian Plate, Russian and English wheats, which were average at the time of purchase and suited the baking requirements of the Birmingham district), wholemeal flour (two-thirds No. 1 Manitobas and one-third English, chiefly reds), ground patna rice and casein of industrial grade.

In the early stages of this experimental work I made drawings and descriptions of the eggs, larvae and pupae of all the species of spider beetles, not only those I was rearing but also others that I was able to obtain, and from them keys for identifications were constructed. In some of the species the sexes were readily differentiated in the adult stage but in the case of the most important species, *Ptinus tectus*, I never found a morphological character which enabled me to tell a male from a female. Right to the end of the work I could only 'sex' this species by dissection. This was accomplished by boiling the dead beetle in caustic-potash solution in a test-tube over the burner of the gas stove in the kitchen until the chitinous integument was softened. The caustic potash was then decanted off and the abdomen of the beetle, which had then parted company with the head and thorax, was washed with water. It was then placed on a microscope slide and with a fine needle the genitalia were extracted. I note from my thesis that I 'sexed' 1,493 *Ptinus tectus* beetles and found 46·8 per cent were females and 53·2 per cent were males. I could only do this, of course, at the end of an experiment. When I wanted some pairs for infesting cultures, cocoons containing pupae were isolated in small tubes and when the adults emerged, which were then, of course, virgin, I placed them together and watched their courtship behaviour. The individuals that tapped quiescent ones with their antennae and made attempts at mounting I took as males and the quiescent ones as females, and I was nearly, but not always, right.

For nearly three years I kept my two ovens going and full of cultures and gradually built up a mass of information. With *Ptinus tectus* I obtained some remarkable results, in that I was able to rear some larvae in pure potato starch which is practically pure carbohydrate, and also in pure casein which in turn is pure phosphoprotein. In both cases I only had to add a very small amount of yeast powder or wheat germ to get a very short life cycle. In the case of casein with added yeast powder the average length of life cycle was only eight weeks.

One of the bits of apparatus I designed was a special sieve—it was my pride and joy. This sieved the foodstuff from the culture tubes at regular intervals, so that I was able to keep a check on the presence of larvae, cocoons and adult beetles in the cultures. I must pay tribute here to my employers of that time, the Dunlop Rubber Company, and especially to the manager of the department in which I then worked, the Special Products Technical. During the daytime I was working on the development of adhesion of rubber to metal, which was then not understood but was vital to the success of the rubber valve of the modern motor tube. I never went in vain to my manager for help. He gave permission for my secretary to type my thesis and the firm's photographer to do the copying of my drawings. My manager helped me in a number of other ways, not least was just simple encouragement, without saying very much. With a man like him at the head of Special Products Technical, where chemists, physicists and mathematicians were employed, there was an extremely high degree of teamwork and consequently much cross-fertilization of ideas, so that the whole was greater than the sum of the parts.

Although as a biologist I was considered to be a bit off-beat, help and advice was freely given to me and as the range of skills available in this great industrial laboratory was exceedingly wide, I can count myself very fortunate indeed and in no wit inferior in this regard to a research student working in a university department.

In the end I wrote up all my results and presented a thesis of 126 pages with thirty-six line-drawings of all the spider beetles I found in the various life stages. In addition I wrote a short section on the various species of mites I encountered in the foodstuffs. They were especially common at the high humidities. I bound a number of copies of the completed work and sent it in. In due course I was summoned to an oral examination by Dr. O. W. Richards and Professor Eastham, which was held in the council room of the Royal Entomological Society, a room I was to know very well in later years. They dealt very kindly with me and finally in 1940 I was informed that I had satisfied my examiners and that the degree of Ph.D. would be conferred upon me.

Sometimes now, when reading through research papers produced by workers in a university that include tables of daily observation, I cannot suppress a sardonic chuckle when I see the observations grouped in fives (Monday to Friday) and then two days missed (Saturday and Sunday) and sometimes four weeks missed with the note: 'It was not possible to continue the observations during the following month.' It could have been put more realistically as 'Easter intervened'.

I have always been conscious of 'time' being a most valuable commodity as far as I was concerned. I had to spend a long time over study. My brother Kenneth had only to read a chemistry textbook o[illegible] and he could retain in his mind everything of value that he required. He did not have to stew it over time and time again as I did. I find the same with writing. The right form of words to express my thoughts adequately does not come quickly, so that I must expend a disproportionately large part of my time not only in clarifying the thought and dividing it into its separate parts but in deciding the appropriate words with which to express it. When I read of the numbers of thousands of words which a professional novelist can put down in the course of a day it depresses me considerably. But I console myself with the suggestion that it does not matter very much what individual words a novelist uses so long as his (or her) plot comes right. In the scientist's case it is very different and certainly in the case of the biologist describing something that must be identified with certainty from all the other myriad forms of life.

Before we had married we had bought a tandem bicycle. This was a very popular form of cycling in the days when the cycle was at its zenith. It was a twin-tube New Hudson with derailleur gears and hub brakes. We had a great deal of fun with it, as it enabled us to get to the forest a little more easily. The added weight of the tandem, however, made it virtually impossible to get it along the Dowles track past Knowle Mill. We entered the forest, therefore, at the Far Forest end, climbing out of Bewdley up Welch Gate, the Long Bank and Callow Hill. We camped on the forest farm of the Bedwells, in a rough sloping patch of ground, partly oak and birch scrub, partly bracken and partly grass. To see and listen to the night-jar, however, was rare refreshment at that time, as my Ph.D. research project, my teaching at the Tech, my caddis-fly biology research and the growing importance of work I was doing at Dunlop on the development of aircraft components involving adhesion of rubber to steel, meant longer and more tiring hours. But, although such lotus-eating was less frequent than hitherto, I was realizing that it was of importance to me psychologically: visits to the forest were indeed not a luxury, they were a necessity.

I would search the flowers of *Umbelliferae* in June or July for longhorn beetles along the path by Dowles or up the firebreak over Brandwood Coppice and would be sure to find the flower-haunting spider, *Misumena calycina*. This is one of the crab-spiders. If disturbed it immediately reacts by holding its front pairs of legs outwards and upwards just as a crab menaces with its claws. The two hinder pairs of legs are much shorter than the front pairs. Its eight eyes are situated around the periphery of a boat-shaped area. It spins practically no silk, so that it does not trap its prey in a web but takes up a position on a flower and awaits the arrival of an insect: a butterfly, fly or a bee. To this end it is able to display adaptive colouration and the extent to which it is able to do this obviously excites no suspicion in the insect visitor. On hedge parsley and earthnut the bulbous abdomen is milky white with a tinge of green and looks very much like a petal in shape, as well as colour, there being no pattern or marking on it with the exception of a few dimples on the smooth surface. When on the yellow-lime-green flowers of sun spurge, which edge the forest tracks in June, *Misumena calycina* is brilliant yellow with lime-green shading, and it is when on this plant that its colour is the best match and the shape of the abdomen is so like the involucre which cups the cyathium.

On hot autumn afternoons I would see the *Rhipiphori* flying over the rough grass near the hedges of the forest-edged fields. The characteristic hovering flight gives the impression of an insect much larger than its half-inch length, but perhaps its spidery legs help to give the impression of grandeur. This beetle, now called *Metoecus paradoxus* (L), has been known for about 200 years to be a parasite of the wasp, living in the wasp's nest and finally emerging to fly in the sun. On the strength of an observation, made in 1890 by Dr. T. Algernon Chapman of Hereford, that imprisoned female specimens laid their eggs deep in the crevices of partly rotten wood (their long ovipositors poking the eggs as far as they could get them in the crack), it is supposed that in nature the eggs are laid on the wooden posts selected by wasps for the raw material of their paper nests. Presumably the small larvae that have sheltered in the cradle crevice all the winter emerge on to the surface of the post in the warming days of spring, and there in some way are able to seize on to the body or appendages of a wasp and to cling to their host during the flight to the nest. This may be similar to the method adopted by the 'triungulin' larvae of the oil beetle *Melöe* which cluster in the dandelion flowers in early summer clutching andrenid bees with their three-clawed toes and get transported to the bee's nests. But, so far as I know, no one has ever found either a newly emerged or wasp-seeking young larva on a wooden post.

But what arguments raged in those pages of *The Entomologist's Monthly Magazine*! Thank goodness the paper was thicker than is now used or surely it would not have stood up to the critical comments of those amateur entomologists of Victorian days, who were keener to score off another author than to fill in the blank spaces in their own observations. Some, including Miss Eleanor Ormerod, the well-known entomologist, thought that the eggs of the beetle were laid in the nest, but hatching was delayed until the wasp larva was ready to pupate, when the young *Rhipiphorus* consumed the complete larva within forty-eight hours.

In the quiet places, where the sun dapples the forest floor like a fallow's back, the *Dilta* jump and run to cover when they are disturbed amongst the dead oak leaves and mossy stumps. Over the scale of time, the few hundred years that coppicing of oak has been practised on these stony slopes is as a few hours ago when compared with the ages that *Dilta* and its closely related species has been on the earth.

Almost certainly the bristle-tails (or THYSANURA) are the oldest surviving insects and represent an evolutionary stage in insect development before wings had been evolved. The possession of wings is so characteristic of insects that one normally carries a picture of wings in the mind when thinking of insects, the colourful and intricately patterned wings of butterflies, the diaphanous and net-veined wings of dragonflies and the buzzing wings of bees. But when forms like *Dilta*

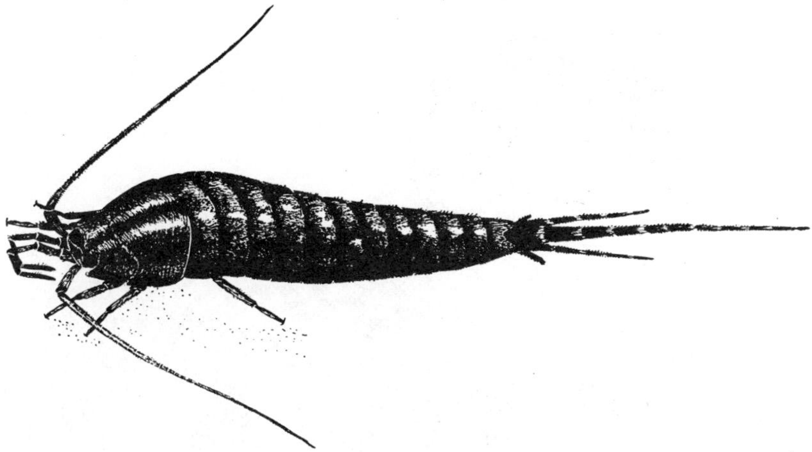

One never seems to find Dilta *when it is the object of a special search, but then it turns up when one is looking for larvae of* Enoicyla pusilla, *the terrestial caddis fly, or when moving stones or house bricks around the Newalls. It is rather patchily covered with fine golden scales but has a formidable appearance when the head is observed under a X10 lens. The compound eyes are joined and the head has a helmeted appearance.*

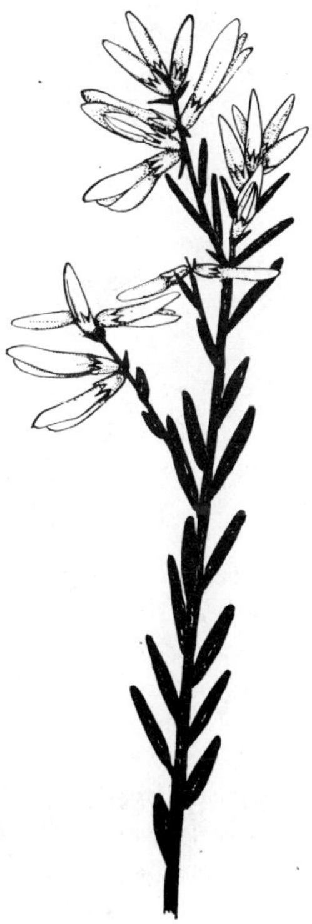

The dwarf broom, Genista tinctoria, *grows in some of the rough patches and along road-side verges on the west side of the forest. Other names for it are Dyers Greenweed and Woadwaxen. The clear yellow pea-flowers open in the middle of July. This plant was the source of a yellow dye which, when used with alum as a mordant and blue woad, gave a fine green called Kendal green. Did not Falstaff complain of his treatment at the hands of 'three misbegotten knaves in Kendal Green'? I cannot think what it is about this plant that interests me, but whenever and wherever I see it I stop and point it out 'Look! Dyers Greenwood!'*

The parasol mushroom, Lepiota procera, *occurs in great profusion in a large field to the north of the Newalls in Wyre Forest. Only a few years ago this part of the forest was felled and cleared. It is probable that the mycelium of this fungus is feeding on the dead roots of trees or other dead organic matter associated with the trees. Books say that the parasol is found in clearings or at the edge of woods. From August to September we could always see about five hundred of the large eight-inch spotted plates, but twenty was a meal for all of us. Do not cook the stems as they are tough and the parasols require more cooking than cultivated mushrooms. There is no need to skin the tops; just scrape off the loose scaly parts. The flavour is delicious but it is possible to have a surfeit. We have not succeeded in drying them.*

Unfortunately, the 'Shaggy Ink Cap' (Coprinus comatus) *season runs concurrently with the parasol and although the flavour of the former is stronger, on the whole they are to be preferred. The difficulty with the ink caps is the speed with which they literally digest themselves. Within a few hours an ink cap can be a pool of black liquid and it is quite impossible to do anything with it in the kitchen. However, deep-freezing halts the process, but tie them securely in a polythene bag or they rapidly take up other flavours in the 'fridge'.*

first scurried amongst the forest detritus the horsetails grew like trees and this was long before any flowering plants had stretched their stems towards the sun. Strange that no oak trees existed when *Dilta* first appeared, but the remnants of the horsetails lie a few feet below the surface as thin coal seams only laid bare when new forest roads are cut as firebreaks. *Dilta* looks more like a shrimp than an insect, especially when looked at sideways, and is more reminiscent of its own ancestor, the trilobite, than one of the more highly developed insects.

I can even now follow the heavily rutted tracks made by the laden timber wagons, deep and narrow with the distance between kept clean of vegetation by the scraping axles themselves. Although the ruts remain, the surface is covered with heather and the silver birch has seeded itself in patches, so that although one can follow along the heather easily, an almost impenetrable barrier of the closely growing young trees is encountered from time to time. In the steep valleys the

In Britain, the number of insects which can attack or bore into sound wood is very small. But as the wood becomes progressively rotted by fungi, more and more insect species of a wider range of orders find their biological environment.

The larvae of Lyctus *beetles will develop only in the fresh sapwood of the oak, soon after the free falls when the starch content is high. When the starch has been respired the female* Lyctus *is no longer interested in it sufficiently to lay her long slender eggs in the vessels. If the bark of an oak tree has been wounded, perhaps removed by cattle rubbing against it, the underlying sapwood is killed and the little* Anobium punctatum, *the common woodworm of our homes, will infest the patch of dead sapwood not as yet decayed except superficially by soft-rot fungal species. But progressive insect attack facilitates heavier and heavier attack by the wood-rotting fungi which again allows further infestation by different groups of beetle species, such as for example the death watch beetle,* Xestobium rufovillosum. *Even heavier fungal attack brings the large stag beetle,* Lucanus cervus, *to lay her eggs and after a number of years a few generations of this species leaves large tunnels in the now soft rot. It is at this stage that some other orders of insects become involved in wood decay. Mostly beetles have, so far, played their part as the insect factor in wood decay but now the wood-boring crane fly,* Tanyptera atrata, *can take a hand. This large black tipulid with the orange band and large upwards-curving ovipositor can be seen around the forest laying her eggs in the rotting oak stumps.*

paths are of two kinds: those that have been worn down to the buff sandstone that peels away in large flat slabs, and those where the clay was so deep that in order to get the oak lumber out of the forest a road had to be constructed. This was done by laying straight oak cordwood transverse to the length of the track side by side. A vast amount of oak branch wood was used in this way. Perhaps the best example that can now be inspected, preserved as like a monument, I remember being made about 1921 to get the wagons up from Dowles Brook at Rudds Bridge past the Newalls up to the metalled road by Wyre Forest railway station. The section that rises steeply from the brook is a wooden road in perfect repair. The oak is soaking wet but steel hard. From the Newalls to the road is another steep rise, but this is solid stone. The lower track is narrow and the hedges overgrown, but what a scene of activity when six muscular and sweating horses were chained to a great log on four wooden wheels. Jessie Carter would shout to each horse in turn to tauten the chains and then command them to pull and the great wheels turned with the drum-sized boss hubs touching the banks on either side.

6

Caddis Pleasure

DURING the last year of working for my degree, when I was steadily ploughing through all the practical work in Balfour-Browne's book, I made up my mind that when I got my B.Sc I would take up some entomological research work, to do outside ordinary business hours and endeavour to get a higher degree. When I succeeded in 1936 I already had some general ideas about the work I wanted to do. The idea of collecting the immature stages of the different species of a group of insects during the weekends and holidays appealed to me, and I wrote to N. D. Riley who was the keeper in the Department of Entomology in the British Museum (Natural History). I explained to him the sort of thing I wanted to do and asked him whether he could advise me about a subject. His reply had a remarkable influence on me.

Department of Entomology
British Museum (Natural History),
Cromwell Road,
London, S.W.7.
January 15, 1937.

Dear Mr. Hickin,

. . . There is, however, one matter which is of some urgency which you might consider, but its pursuit depends very largely upon how you are situated and what facilities you have for field work. As you probably know, there has been a certain amount of work done in recent years upon the ecology of fresh water lakes and streams. One of the predominating entomological factors in these situations are the caddis flies. The adults of these creatures are now identifiable with certainty owing largely to recent work on their genitalia, but it is still extremely difficult, if not impossible, to identify with certainty many of their larvae. Several recent papers have included lists of the Trichoptera in which the identifications of the larvae have been made with the aid of keys based on the continental fauna, and they have resulted in the recording as British of several species which, as far as we know, do not occur in the British Isles at all.

You will see that the work therefore entails not only detailed morphological studies of the larvae, but also the rearing of these to maturity so that their identity can be checked. The ultimate aim would be to publish detailed accounts of the larvae of all our British species, and, if possible, keys to their identifications. If you can achieve this, you would be performing a lasting service, not only to entomologists but also to the fishery people, but the task would not be simple, and would almost certainly occupy all your spare time for several years. If you feel inclined to take it up you may rely upon co-operation in so far as it is possible to give it.

Believe me,
Yours sincerely,
(Signed) N. D. Riley,
Keeper of Entomology

His suggestion was that I should make a study of the immature stages of caddis flies, the insect order known to entomologists as the TRICHOPTERA. Riley explained that although the adult stages of the British caddis flies were well known, next to nothing was known of the larval stage. Most people knew what caddis grubs looked like, but we had no detailed standard work on the immature stages of our British species. It would be a lifetime's work and a vast number of time-consuming drawings would have to be made. Riley suggested that Martin Mosely, who worked in the Entomology Department of the Natural History Museum on the caddis flies, would gladly give me all the help I needed.

I immediately set about carrying out this suggestion and got in touch with Martin Mosely. He was just publishing his own book on the TRICHOPTERA, but although a classic for the identification of the adult insects, it contain next to nothing concerning the life cycles and immature stages. I was very grateful for the time and trouble he took with me and he saw my papers being published before he died.

The idea, as detailed by Riley's letter, was that I should study the immature stages of the TRICHOPTERA and describe them in a series of papers. Very little had been done on this group, although they were very common insects around all fresh water from ditches and streams to ponds and lakes. Perhaps I should at this point say something about the caddis flies by way of general background, because they have been so important to me from that time onwards.

Caddis flies are moth-like insects, but usually with more slender wings, and instead of the powder-like scales which cover the wings of moths, caddis flies have rather hairy wings, which gives rise to the name of their scientific order, the TRICHOPTERA, which literally means hairy-winged. In some species, however, the hairs are flattened, often elaborately. In size they vary from moderate to very small, but they never approach the size of the larger moths. The point perhaps of greatest interest concerning them is that their immature stages are aquatic. Only one genus, *Enoicyla*, shows an exception to this and there will be more about this later. There is only one exception to the rule that they are fresh-water insects; the larval stage of a species which is found in brackish water in New Zealand.

Caddis flies are usually to be found near water and they are commonly seen flying over ponds and rivers in the summer evenings. They are usually drab in colour, but there are exceptions to this. There are nearly 200 different species, classified into thirteen families, recorded as occurring in Britain, and it is certain that a few more species will be found. I myself found one, *Tinodes pallidula*, in Surrey, which McLachlan, the great British collector of this group and allied orders, had described from specimens he had collected in Bulgaria. In this case I collected the larvae first and reared them out, and as far as I know the only British specimens are from my aquaria.

I determined rather early on in my studies that I would not make a 'set' collection of the adult caddis flies, as this takes up rather a long time, setting the insects out on to boards as butterflies are set. I thought it would be much better use of my available time to do the maximum amount of illustrating with drawings, much of this being extremely time-consuming, although rewarding in other ways. This has meant, however, that without a standard set and named collection it has often been difficult to identify the adult insects when I have reared them, but I have been fortunate to have the very willing co-operation, since Martin Mosely's death, of Mr. D. E. Kimmins, who took over the responsibility for the study of this group. For over twenty years D. E. Kimmins has identified my adult caddis flies, not only those I have reared in England but also the large collections I have made during my visits to East Africa.

The female caddis fly lays her eggs on a water plant at the surface of the water, but some species are known to crawl under water in order to lay their eggs on stones at the stream bottom. The eggs are laid in a single egg mass which when coming into contact with water swells up rapidly. In an aquarium, or just a jam jar, one can watch the little grubs developing inside the eggs and then after about ten days the grubs emerge from the egg mass and crawl about in the water.

Thereafter, what they do depends on what sort of caddis they are. A large number of species construct cases from which only the head and legs protrude. These cases are often characteristic of the species of caddis grubs which makes them. For instance, one sort likes to make its case of the shells of water snails, whilst several construct their cases of sand grains with or without small pebbles at the sides. Another beautiful case is made from circular discs cut from leaves which have dropped into the pond from trees. These discs are geometrically accurate, far more accurate than a human could cut armed with a pair of scissors. Another species makes its case from sand grains but attaches one or several fairly large twigs on the side. This is probably the origin of the name, 'Stick worm', given to them in some country districts. The beauty, the accuracy and the characteristic patterns shown by the cases made by these insects in our ponds and streams have been a source of wonder to biologists over the past few centuries.

The caddis grub is not a delicate feeder, some are predominantly vegetarian, eating all sorts of aquatic plants and vegetable debris, others, on the other hand, are mainly carnivorous, catching small water insects and devouring them, but most species will eat each other on occasion. When they are fully grown the entrance to the case is blocked up either with a small pebble, a plug of vegetable rubbish or with a net made of silk and after a short resting period the grub changes into a pupa or chrysalis. This is much like the adult insect except all the parts are soft and yellowish white and the wings are short and tucked in round the sides of the body. It usually only remains in this stage for about a couple of weeks, then the pupa cuts its way out of the case by means of a pair of saw-like jaws it is furnished with and swims vigorously around with its specially fringed middle legs until it comes to a reed or a stone which breaks the surface of the water. It crawls out into the air on this, then metamorphoses into the adult caddis fly. A split appears down the middle of the back and by a series of rhythmical pulsating movements the caddis fly just emerges from out of its old pupal skin. The wings expand to their full extent and within a few minutes (sometimes within a few seconds) the adult is able to fly or crawl away. Some species of caddis can accomplish their final metamorphoses on the surface of the water. It very often happens that thousands of caddis emerge together, then the 'hatch' is on, and if the water supports any trout it almost boils with their activity.

Not all caddis larvae bear cases, although this is the popular conception of their habits. A number of species in three families construct bag-like nets into which are swept by water currents small organisms which are captured and eaten. These larvae are very agile and able to move more quickly than those which have to drag their case along with them. They are easily found by moving a few stones from the bottom of a rapidly flowing brook, where the nets are easily seen. Nets of some species, however, are found under leaves of water plants, and I once bred a large number of the very rare net-spinning species, *Holocentropus stagnalis*, from the floating liverwort *Riccia fluitans* in an aquarium where I was rearing the large smoky-winged and yellow-necked *Neuronia ruficrus*. I had no idea the little *Holocentropus* were in the aquarium until they were flying in the room. I have been back to the little woodland pond in Worcestershire many times but I have never found the larvae.

Another family of caddis have larvae which live in long silken tunnels on stones or sunken sticks and I have had great success in collecting the different species of this group. One species spins its long tube on wet rock and I cannot pass a slimy wet rock surface without examining it to see whether it is there. Incidentally, it always intrigues

Few caddis flies possess the brilliance of colouring of the butterflies. Usually their wings are decorated only in browns, blacks, cream and white, but the broad-wings of Neuronia clathrata *are yellow with the dark brown reticulations of the veins. The yellow is not a canary yellow of the butterflies, but nevertheless this very rare insect is of remarkable appearance.*

I have searched for its larvae in vain at Chartley Moss, one of the only two localities where it is known to occur. This is a small lost piece of England in the heart of Staffordshire, a floating raft of sphagnum moss and the only place where I have seen trees quiver and move when I gingerly walk twelve feet away from them.

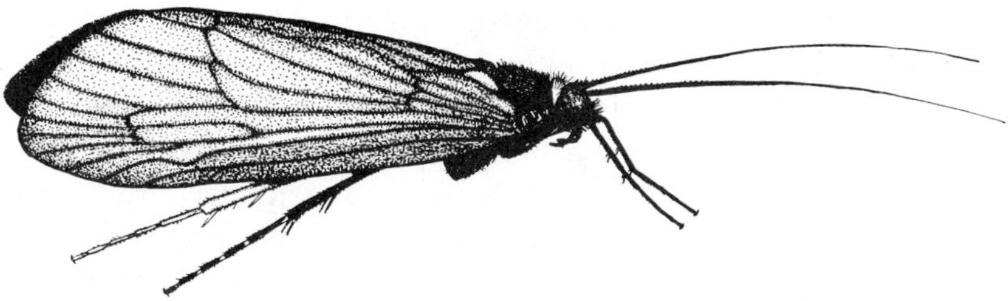

Neuronia ruficrus *was the first rare caddis fly that I reared from larvae which I collected at Broadmoor Wood, just inside Worcestershire at Bromsgrove. It is smoky black but in the living insect the neck is bright yellow.*

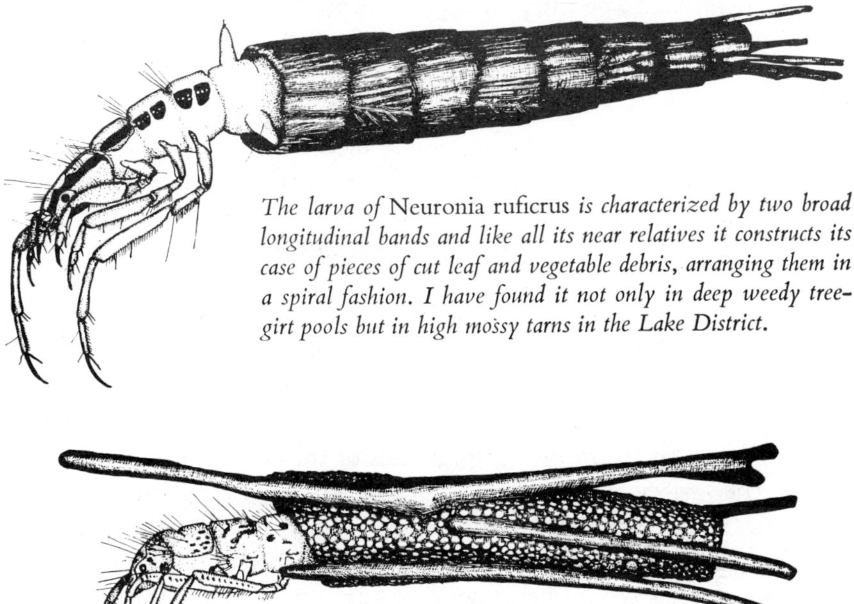

The larva of Neuronia ruficrus *is characterized by two broad longitudinal bands and like all its near relatives it constructs its case of pieces of cut leaf and vegetable debris, arranging them in a spiral fashion. I have found it not only in deep weedy tree-girt pools but in high mossy tarns in the Lake District.*

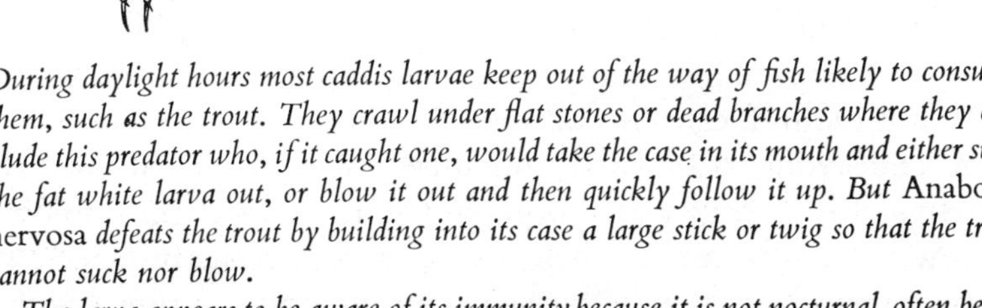

During daylight hours most caddis larvae keep out of the way of fish likely to consume them, such as the trout. They crawl under flat stones or dead branches where they can elude this predator who, if it caught one, would take the case in its mouth and either suck the fat white larva out, or blow it out and then quickly follow it up. But Anabolia nervosa *defeats the trout by building into its case a large stick or twig so that the trout cannot suck nor blow.*

The larva appears to be aware of its immunity because it is not nocturnal, often being found in considerable numbers walking over sandy spits in the stream bed in broad daylight.

me to use the name which biologists have conjured up for the animal life found on a wet rock surface, the *fauna hygropetrica.*

Caddis flies have a great interest for trout fishermen as in all stages they are eagerly sought after for food by the fish. This has resulted in two fishing methods, known as wet-fly and dry-fly fishing. In wet-fly fishing a model of a caddis pupa swimming to the surface is used, complete with hook, and in dry-fly fishing the artificial fly is made (in many cases) to mimic the appearance of a caddis fly or a may fly, although some of the fishermen's artificial flies are just imaginative bits of coloured fur and feather.

Most of the caddis flies are known to the biologist or entomologist only by their scientific names, but a number of species have been known for upwards of a century by their fisherman's names, and what delightful names they are. We must perpetuate them. Here is a short list of them:

The Cinnamon Sedge	*Limnephilus lunatus*
The Welshmans Button	*Sericostoma personatum*
The Grannom	*Brachycentrus subnubilus*
The Black Silverhorns	*Leptocerus aterrimus*
The Grouse Wing	*Philopotamus montanus*
The Grey Sedge	*Odontocerum albicorne*
The Large Red Sedge	*Phryganea striata* or *P. grandis*
The Sand Fly	*Rhyacophila dorsalis*

One of the most pleasing features of the collecting side of the work was the variety of very pleasant localities where I searched for specimens. Sometimes I collected 'blind'; that is, having found a suitable place where caddis larvae would be found, I searched for them not knowing which species would be present, at least until I knew them well, when I would almost instinctively know which species of caddis would be present. The other method of collecting was to read up the old notes in entomological literature in which localities were given where species of adult caddis flies were collected. I would then search out the locality and endeavour to find the immature stages, if indeed they were still present. Much help was also given to me by local entomologists who knew the caddis flies of their district and in some instances, after my first papers had been published, entomologists wrote to me saying that they had found such a species, would I like to come and find the larvae and they would render all assistance. In this way I explored Askham Bog, Malham Tarn in Yorkshire, many lovely places in the Lake District, Chartley Moss in Staffordshire, many of the little streams and pools in the half-forgotten side of the Cotswolds, Cannop Pool in the Forest of Dean and many other places besides. Sometimes I found the species I was looking for as well as additional species, at other times I was unsuccessful, but always the localities, being beside water, running streams, lakesides or rivers, were very pleasant places.

On the first occasion I met Martin Mosely at the British Museum he made a list of the principal books on the TRICHOPTERA that I would require to have at my right hand. These included Ulmer's volume in the *Süsswasserfauna Deutschlands* and *Les Larves et Nymphs Aquatiques des Insectes d'Europe* by Rousseau. These cost £2 and £8 respectively. This was during my first year of married life when such sums were enormous by comparison with my salary, but Emma encouraged me to get them, as she did all my books, and they have been well-used since. My library on the world's caddis flies began thus.

My first twenty papers were published during the war years, but in spite of this I had an immediate response from entomologists and hydrobiologists for copies of my papers in exchange for separates of their own works. This built up my own library on this subject fairly rapidly, as in most cases I received far more than I was able to make by way of return. Immediately after the war a number of European entomologists got into touch with me. Among them was Ulmer, the grand old specialist on the group, and he presented me with copies of all his papers, a considerable collection. He had his first paper published on caddis flies in 1909. I visited him at his home near Hamburg and he delighted in showing me his great library of world literature on the TRICHOPTERA and his vast collections. He was a man of single-minded purpose in his research specialization, living a very simple bachelor existence with a very strict routine. Physically, everything was against him, as he was dwarf-like, hunch-backed and had a heart affliction, but until he died in 1962 he never failed to write me an appreciative and encouraging letter acknowledging each of my papers.

When I had drawn a series of four larvae in such detail that I believed they could be identified from all the known British species, I took the folio up to Dr. E. B. Worthington, who was then the director of the laboratory of the Freshwater Biological Association which was situated at Wray Castle on the side of Lake Windermere.

When he examined my work he pointed out that I was sure to meet difficulties in reproducing the illustrations because of the style of shading I had used. Although I had drawn a firm black outline I had used pencil shading. He kindly found me a book in his library by Professor Graham Cannon that gave instruction in using black dots for shading, placing them closer together where the shadow was to be deeper, but larger and further-apart dots or smaller dots, where the shadow was least. I took my drawings home and changed the shading to the black dots, and until I started to use the scraperboard technique I followed this method entirely. I tremble to think how many dots I have put on bristol board since that time. I followed this practice because the dots cannot be mistaken for anything else, whereas if lines are used for shading they may be mistaken for hairs or bristles or may even give a mistaken impression of the shape of the organ or the part of the insect being depicted.

I also had to familiarize myself with the technique of making the drawings at least twice as large as they would ultimately appear on the printed page and I had some difficulty with details being lost when my first drawings were published.

One of the advantageous features of doing this sort of work, especially as it has always been entirely a leisure-time occupation, is that very little is required in the way of equipment. For rearing the caddis larvae I have used almost every sort of container that would hold water, from enamel pie dishes and jam jars to glass aquaria made from battery jars. For the pond and lake-dwelling species a few aquatic plants are tied to an alga-covered stone and put into the dish, together with some sand. I once introduced some more caddis larvae of another species accidentally in the weed, so that this possibility must always be examined carefully.

Larvae which have been collected from swiftly running water are a little more difficult to cope with, as they not only require a degree of oxygenation, which can only be provided artifically, but they require the water to have substantial movement. Some of these species are dependent upon small organisms being swept into their webs or nets or probably directly into the brush-like organs around the mouth. This can easily be accomplished by the simple oxygenating equipment sold for tropical-fish aquaria and stirrers can be rigged up to provide the necessary degree of water movement.

A stick or a cork must also be placed on the water so that when the swimming pupa comes to the surface for the final metamorphosis into the caddis fly it has something on to which to crawl. The top of the dish must be covered with muslin fastened with rubber bands, in order to prevent the escape of the caddis after all the time and trouble spent in rearing it. All my research specimens are preserved in a mixture of alcohol and formalin together with their associated larvae, pupae, cast skins, etc. Full data must always be written on pieces of paper *inside* the bottles containing the specimens so that they can never fall off and be lost or mixed up. By data I mean the locality and county where the specimens were collected, the date of collection and the name of the collector. Without this information specimens are of no scientific value whatsoever. The way I numbered the specimens in my collection may be of interest. To every 'batch' of caddis larvae that I collected that were obviously of one species I gave a four-figured number, the last two figures of the year of collection becoming the first two of the serial number. For example, the fourteenth batch of larvae of one species collected in 1948 is numbered 4814. Each year I used a new stiff-covered notebook and entered all the details of collection and rearing against each serial number.

When larvae had been reared to adult stage and the identification confirmed by D. E. Kimmins at the British Museum I preserved the adults in alcohol in a small tube. They were then placed in the bottle of preserved larvae from which same original batch they had been reared. When I had made a set of drawings and written the description I gave the species a serial number in my series of papers, 'Larvae of the British TRICHOPTERA', which have all been published in the Proceedings of the Royal Entomological Society. This series contains fifty-two papers describing fifty-six larvae, but in addition there are two general papers, one on the larvae and the other on the pupae of TRICHOPTERA. My descriptions of African caddis larvae are contained in three papers published in the same journals.

When a batch of larvae with its associated adult flies had been used in one of the published descriptions, I placed a large number (coloured blue) in the bottle of specimens, the number being cut from a calendar. Although this may sound rather complicated actually it is very simple and it makes absolutely certain that the specimens are not mixed up, and also enables any research student coming after me and wishing to check my work to find, fairly easily, all the specimens I used for drawing and in making the descriptions.

A number of my drawings have been reproduced abroad in works on TRICHOPTERA and general works on aquatic insects, notably in the United States, France, Belgium, Russia and Rumania, all acknowledged with the exception of the French book which reproduced ten of my drawings.

I feel that I was very fortunate in having such delightful places in which to collect my specimens. Almost all of my early caddis-larvae collecting was firstly with Emma and then later in addition with one, and later still, with two daughters whom we named Verney and Sari. At the end of the war I took an appointment as an entomologist with a company manufacturing insecticides and we moved down to the village of Bletchingley in Surrey where we lived for ten years. We were able only to walk out of our house, pushing the children in a pushchair, to reach a little stream where we could picnic and I could collect. This was where I found the species already referred to, *Tinodes pallidula*, for the first time in Britain. Indeed, the finding of larvae of fresh species hitherto often unknown to me, was a family affair.

I must tell the story of our finding the larvae of the species *Enoicyla pusilla*. This species is the only one in the British list whose larvae are not aquatic. They were said to be found in moss at the foot of trees and the only records of their being found in England were nearly a hundred years ago in the county of Worcester. For many years I had been searching for them in Wyre Forest without success and I had been in correspondence with entomologists on the Continent who had described in great detail the sort of habitat in which they would be found. Then at Whitsun 1958 I had been camping in Wyre Forest with my younger daughter Sari, and we were just packing up our tents when Sari exclaimed, 'Daddy, I've got a caddis on my tent!' 'Don't be foolish,' I replied, 'it's just a case-bearing clothes moth, I expect.' 'Well,' she said, 'they look like caddis to me and I've got some walking up my leg,' not being put off. So I took one and examined it with the naked eye, and seeing the little tubercles on the first abdominal segment I pronounced it *Enoicyla pusilla*, the terrestrial caddis, rediscovered in Britain after nearly a hundred years. We then found hundreds all around us and we had not only been living with them cheek by jowl that Whitsun but for some years I had been camping amongst them.

One pleasant task I undertook with my study of caddis larvae was the writing of a little book about them. I had met Dr. Eric Ennion at the Flatford Mill Field Study Centre and he suggested that I wrote a

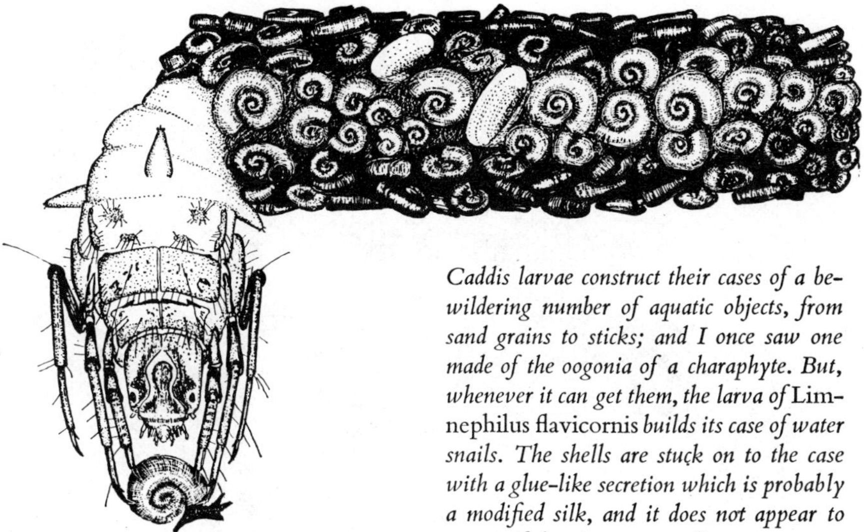

Caddis larvae construct their cases of a bewildering number of aquatic objects, from sand grains to sticks; and I once saw one made of the oogonia of a charaphyte. But, whenever it can get them, the larva of Limnephilus flavicornis *builds its case of water snails. The shells are stuck on to the case with a glue-like secretion which is probably a modified silk, and it does not appear to matter if the contents are alive or not.*

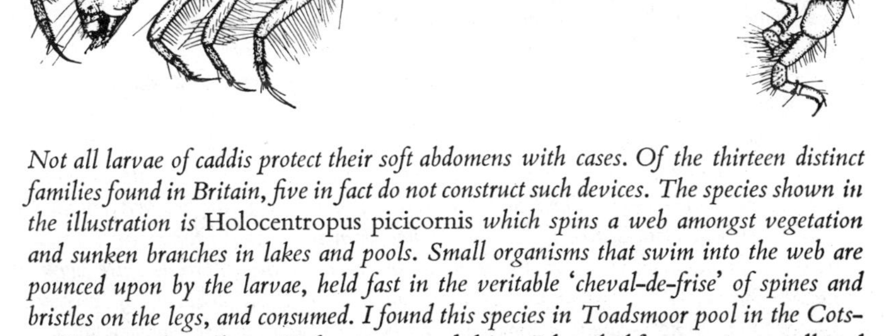

Not all larvae of caddis protect their soft abdomens with cases. Of the thirteen distinct families found in Britain, five in fact do not construct such devices. The species shown in the illustration is Holocentropus picicornis *which spins a web amongst vegetation and sunken branches in lakes and pools. Small organisms that swim into the web are pounced upon by the larvae, held fast in the veritable 'cheval-de-frise' of spines and bristles on the legs, and consumed. I found this species in Toadsmoor pool in the Cotswolds but its close relative,* Holocentropus dubius, *I described from specimens collected from Barnslands near the Cleobury Mortimer edge of the forest.*

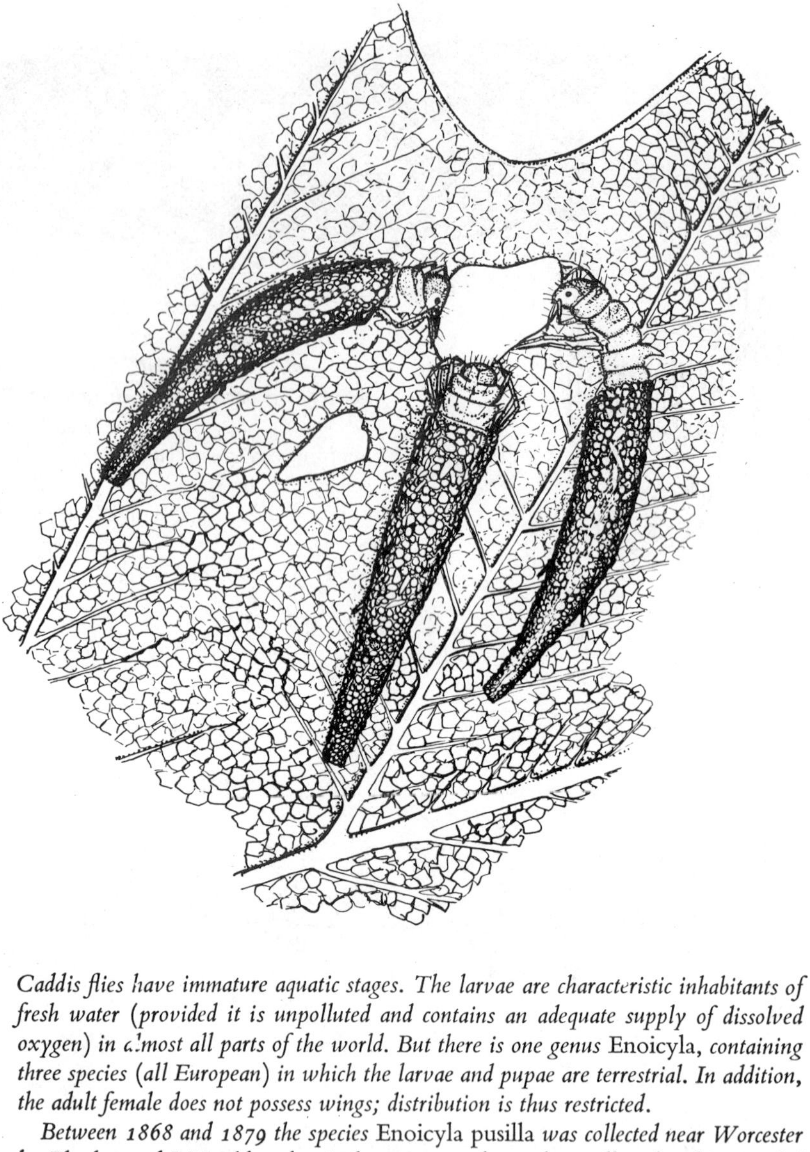

Caddis flies have immature aquatic stages. The larvae are characteristic inhabitants of fresh water (provided it is unpolluted and contains an adequate supply of dissolved oxygen) in almost all parts of the world. But there is one genus Enoicyla, *containing three species (all European) in which the larvae and pupae are terrestrial. In addition, the adult female does not possess wings; distribution is thus restricted.*

Between 1868 and 1879 the species Enoicyla pusilla *was collected near Worcester by Fletcher and McLachlan, but it then went unobserved, uncollected and unrecorded until April 22nd, 1957, when Sari found a few larvae walking on her tent in the Forest. Although our search for* Enoicyla *had been unsuccessful for many years, we suddenly found them in great profusion in the area on the Worcestershire bank of Dowles Brook. Within a few yards of the original site of Sari's discovery of* Enoicyla, *Emma found 'The Newalls' which now fits perfectly into our arrangement for forest refreshment.*

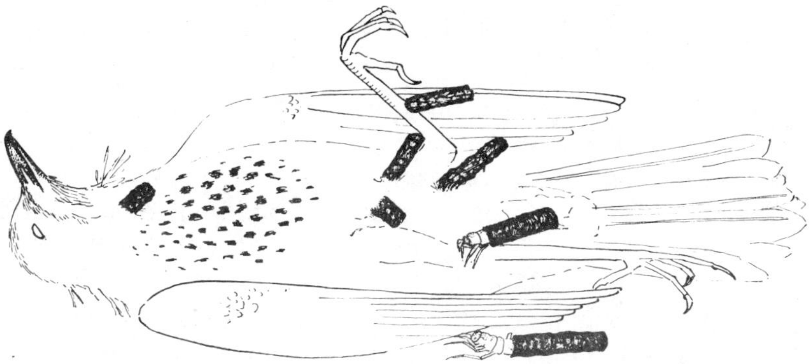

During the course of my work on the biology of caddis flies I found out what I thought was a remarkable thing. Furred and feathered animals whose dead bodies got into streams and ponds were consumed by caddis larvae. No one had ever recorded this before. I first observed this near the Forest of Dean where, in a ditch I was examining, I found a large writhing mass of caddis cases completely covering the dead body of a mouse. Then I found a dead coot in Malham Tarn in Yorkshire on which a number of caddis were devouring the flesh. In 1962, I found a body of a song thrush in a little stream that runs into Dowles and a number of caddis larvae were half hidden in the feathers.

little book in the Field Study series which Methuen were publishing. This did not take me long, as it only had fifty pages and forty-seven illustrations, and the Royal Entomological Society gave me permission to reproduce some from my research papers. I was very short of space, but I hit upon the idea of using the end papers for describing the morphology of caddis larvae. By this means I managed to get four more pages which I arranged like the pages of a laboratory notebook with fully labelled drawings. This gave me twenty-eight extra drawings taking the place of a glossary of scientific names. No one had ever done this in a book before. There are four colour plates and Methuen made a very beautiful book of it which they published for 9*s.* 6*d.* and I have often had compliments passed to me concerning it. Methuen's paid me a lump sum for the book, which represents the only remuneration I have ever had in connection with caddis-larvae research, except the travelling bursary from the Royal Society and the Visiting Scientist appointment to the East African Fisheries which I write about in a later chapter.

* * * * *

What part has the forest played in this work which has given me so much pleasure during the last twenty-seven years? In the first few years of the work before the war I did a certain amount of collecting and, indeed, I see that in the first twelve descriptions four of the species I collected were from Dowles Brook and another two species were from the Cleobury Mortimer district nearby. In addition, of course, the terrestrial caddis is found only in the Forest as far as Britain is concerned.

Although my caddis-research work has slowed down during the last few years, I am desperately keen to continue it on an even larger scale, which I must do if I am to reach anywhere near finality in describing and drawing the British caddis larvae. I have only completed a quarter of the known species, and although other investigators have described a number of others, the greater part of this work is still to be done. I feel that the forest will give me caddis-pleasure in the future when I am able to explore caddis-wise every brook and every little rill that runs into it.

7

Away from the Forest

CLEOBURY MORTIMER is a little Shropshire town lying about a mile to the west of the Forest. From the village of Far Forest, lying on the Worcestershire brow of Lem Hill, the road plunges downwards to the brook that divides the counties. It then winds through the outlying woods of the Forest to the extreme west, Bell Coppice, Hungry Hill, Coachroad Coppice and Prizeley, for just over two miles before twisting round the entrance to Mawley Hall, until Cleobury is sited. At this spot when I was a boy an old cherry orchard lay on the high right bank which gave me the breathtaking delight of the meadow saffron. The purple carpet of the autumn crocus has not existed for many years and the cherry trees have gone too, but strangely as I pass this spot today (as happens from time to time) I look around half expecting them still to be there.

It was in Cleobury Mortimer that I found and purchased Spregdon House, a largish brick-faced multi-windowed house fronting on to the High Street. It had no side entrance, just one front door, but there was a small coach house and stable at the bottom of the garden which opened on to a back lane. Spregdon House was in a disreputable state of repair and had few modern conveniences; water heating and cooking being carried out on a large black iron stove in a kitchen also large and black. I papered and painted the whole of the eleven rooms in as bright a décor as could conveniently be managed, and Emma and our two-week-old Verney moved there with my sister-in-law Jessie, Kenneth's wife, with her baby daughter. This was a disastrous move and we were all very unhappy and it was not long before the two girls and their babies were back in Castle Bromwich again. After keeping Spregdon House for some time I sold it at a modest profit, but if I had kept it a little longer a handsome profit would have been made. We kept the stable and coach house and during holidays and a few free weekends I converted it into a little country cottage. We called it Piers Cottage, after the Vision of Piers Ploughman of William Langland, who was reputed to have been born in Cleobury Mortimer in 1332, although he wrote his poem on the Malvern Hills. We enjoyed our visits to Piers Cottage and from it we were able to get to delightful streamside picnics and caddis-collecting places on the river Rea and its tributaries from the north, that rise on the east side of the Clee Hills. I found a large number of the larvae of the caddis fly *Lepidostoma hirtum* at one spot where the river Rea had a wide pebbly beach and little pools off from the main stream. This caddis makes a case of various pieces of vegetable debris that are roughly square in shape and remarkably enough the case is also square in cross-section. A square is a geometric shape rarely achieved by insects in building patterns. My drawing of the larva in its case has been reproduced in textbooks

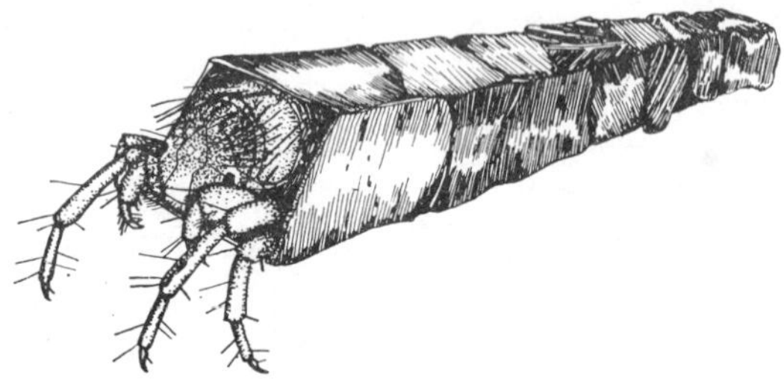

The grubs of caddis flies construct their protective cases in many strange and varied patterns and with a wide variety of materials which they can find in the streams, rivers, ponds and lakes. Mostly, they are round or oval in cross section but a few species construct cases with a square cross section; one such is Lepidostoma hirtum. *I reared a number to the adult stage so that I could be sure of their identification.*

on entomology in a number of foreign countries, and when I see these my mind always returns to the shady bank on the river Rea behind Shakenhurst and where we were rudely interrupted by the landowner as we had strayed a few yards from the footpath.

The weir at Lower Forge is one of the most beautiful river-falls that man has devised, at least that I have set my eyes upon. The water runs down a series of wide rock steps mantled with the dark green lush velvet of *fontinalis* moss in which large numbers of the little caddis

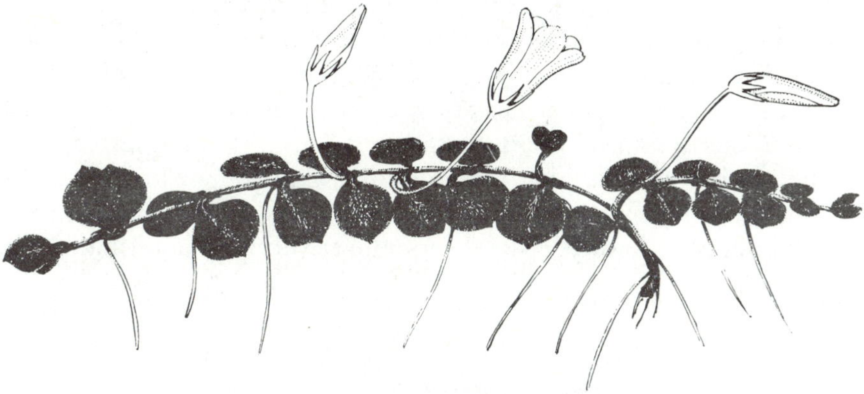

What pen or brush can depict with accuracy the superb delicacy of the bog pimpernel, Anagallis tenella*? Where the sodden sheep-trodden turf seeps into the clear rills, its presence is almost imperceptible. But the little fragile pink bells cluster in the moss and only on the closest inspection can the creeping stalks edged with almost round and seemingly veinless leaves be found. A hundred years ago it was found in the forest but, alas, it is there no more, and the deep-cut valleys of the Long Mynd in Shropshire now provide me with this delightful sight.*

Hydroptila occulta abound. It seems now lost in another world and the derelict dower house of Mawley Hall that stood nearby has now been removed so that one can hardly find a single brick.

My work on adhesion of rubber to metal by the brass plating process led from the motor-car tyre valve to a large number of accessories used in aeroplane manufacture. The first was a device called the Bristol Bush. It consisted of two concentric steel cylinders with rubber moulded and adhering in the intervening space. They were used in Bristol radial aeroplane engines as mountings. This type of product was necessary for the improving performance of aeroplane engines in order to give flexibility, yet at the same time certain dimensions were to be of the order of those required for all-steel parts. There were many problems and I never seemed to be out of trouble in the early stages of the development of these products. After 'Bristol Bushes' came Lord instrument mountings. These consisted of a central steel cylinder held in a perforated steel plate by a web of rubber. There were several sizes and each size was made in a range of loadings. These products were for the mounting of separate instruments and also whole instrument panels in order to insulate their delicate mechanisms against vibration, jolts and jars. Vast quantities of these were made before I went on to another project. I also had the responsibility for making rubber washers and diaphragms to very fine dimensions. These were made in long tubes then cut on a lathe. They subsequently had to pass just as rigorous a dimensional inspection as for steel parts. This was a new development. The old rubber vulcanite, rubber with a high proportion of sulphur, came into its own again for gun-firing grip finishes. This was because it did not crack or otherwise deteriorate at very low temperatures.

When the snow has lain for some weeks, then suddenly melts, the busy work of the bank vole, Evotomys glareolus, *is revealed. The tunnels made of bits of chewed grass lie right on the surface of the turf, stretching from the forest edge banks right into the middle of the old tussocky paddocks. Grass is obviously eaten and small segments of the green blades lie about. Immediately after the exceptionally wintry conditions of February and March 1963, evidence of great activity by the bank vole was everywhere. The blanket of snow gave protection from the Tawny Owl, and freedom from wet or damp and mouldy food.*

Then one day the head of my department sent for me about five in the afternoon and asked me what jobs I was currently engaged on. I made a verbal list but he was not listening to me. When I drew breath he said: 'Hickin, do not spend another minute on any of this work, I am making arrangements for someone to take over from you. In the morning you are going to a factory in Wolverhampton where "self-sealing" is applied to aircraft petrol tanks. You will in future make a study of the "self-sealing" techniques, because this company has been given the responsibility of enlarging by many times the facilities available and I want you to be in charge of it.' The clean break from all the thousands of different aircraft parts with all their worries and problems suited me very well, but I soon realized that I had collected an even larger batch.

The usual method of sealing an aircraft petrol tank, so that when a bullet or piece of shrapnel went through it the petrol was not all lost, was to cover the tank with a material which was caused to swell rapidly by the emerging petrol. This was accomplished by using a thick layer of partly vulcanized sponge rubber as a sandwich between two layers of strong canvas. An aircraft petrol tank is a very complicated piece of equipment, with all sorts of odd bits sticking out here and there, so that the application of the self-sealing layers was a lengthy and tiresome process as various overlaps for the several layers were stipulated and they had to be in certain places, and there were rigorous rules which had to be followed. The first tanks I covered were Blenheims, thereafter Stirlings came along and finally a large quantity of tanks for Lancaster aircraft. Fighter aircraft had their petrol tanks self-sealed by a different process. A very remarkable material called DTC 10 was used, which was made from a rubber latex sponge, but the vulcanization process was not taken to completion, the semi-vulcanized sponge being pressed flat. This material had no grain effect. If a small cut was made and then the cut used to start a tear it did not tear very far, as the grain turned round on itself. Thus if a bullet pierced the covering of a Spitfire petrol tank the hole made by the bullet was very small, just like a pinhole. It stretched, and after the passage of the missile it closed up so that it could not be seen. I have not heard whether any peace-time use for this material was ever found. After a time I was able to see the actual proof of the success of self-sealing by the repaired tanks that came in for self-sealing, tanks that had been pierced but had got home.

I joined the Home Guard on its formation and took turns in guarding the local telephone exchange, reaching the rank of lance-corporal. On the formation of anti-aircraft rocket batteries, I asked for, and was granted, a transfer to the battery at Walmley Ash, which was quite a fair cycle ride away. However, I threw myself into the battery training programme and went through every rank up to and including captain. There was naturally a great deal of secrecy about this form of defence and it was called 'Z'. The idea behind 'Z ack-ack' was the saturation of a given volume of air-flying space with high explosive. There was not sufficient accuracy in aiming to hit the target, but so many rockets were sent up that the laws of probability and chance made certain that the aircraft passing through the selected volume of sky would have a very rough time of it. The full battery consisted of sixty-four projectors, each of which fired two rockets. Many weekends were spent on firing practice, the camps being at various sites around the coast, and special trains were put on to take the battery personnel to them. Firing the rockets was an awe-inspiring experience. When they were fired, which was done electrically, they belched flame and roared off. It always surprised me that one was safe at the side of the projector. But often the rockets did not move off the projection rails when they should have done so and a long misfire drill had to be undergone at the end of each Sunday morning's firing. I once had to deal with twenty misfires at a firing practice on the sea-front at Wallasey. The projectors were laid on the concrete slabs of the pavement, which I thought gave an even greater degree of harshness to the explosive roar of the rocket back ends when they shot away. When at last the second-lieutenant and I found out what was wrong with each one and successfully got them away I put two pints of beer down without wetting the sides. I commanded the battery every eighth night until the stand-down. No doubt the eight relief commanders, our sixty-four officers and our 1,600 men, all of whom did full-time jobs, saved some military power. We certainly frightened the enemy with our noise, because raids by manned aircraft were a rarity after a few 'Z' batteries had sent up a bagful.

Thus it was my boast that during the daytime I tried to keep them up (ours) and at night I tried to get them down (theirs).

The war years were very full. Dunlop hours were often long, especially when a new design of aircraft petrol tank self-sealing was being worked out and when the first set of Stirling aircraft tanks were received we worked on them for forty-eight hours on end without rest. My Technical College teaching continued for a time, but when bombing became a great hazard classes were transferred from evenings to Saturdays and Sundays. This caused great pressure on my time, especially when I commenced to rise through the various ranks of the Home Guard, so that inevitably I had to resign my biology teaching appointment when someone else could be found to carry on.

During this time we did not get very far from Birmingham. Piers Cottage was about twenty-eight miles from home, but except for our visits there the Warwickshire lanes were our only refreshment. Even so, a number of natural history events of importance took place. My mother and father had a caravan near the water mill at Welford-on-Avon and we sometimes cycled there. One evening we were able to watch a family of otters, two adults and two young ones, having a wonderful time in the fast swirling water where it leaves the weir. A fish had been caught and it was constantly being thrown into the air by a parent otter and then retrieved by a young one. The agility and speed could not but be marvelled at. In a little wood adjacent to a pool at Bickenhill, quite accidentally we found the marsh violet. The finding of this rare plant was quite an event and gave me great satisfaction.

I was also able to find some caddis larvae of a species then new to me in a stream that ran across the flat grazing meadows of Temple Farm at Temple Balsall, where Edmund, Emma's brother, worked. It was a little further afield to get to Chartley Moss in Staffordshire by tandem, but this was a great event. This area of floating sphagnum, although now very small, extending to only a few acres, is representative of what large areas of the North Midlands looked like before extensive drainage. A well-known naturalist, now dead, wrote a long and detailed letter giving instructions how to find it, and this we did. I did not find the larvae of *Neuronia clathrata*, which is found in Britain in only one other locality, but I found the immature stages of two other rare species there.

In 1942 I bought my first motor-car, a 1936 Ford 8, and I paid £50 for it, and when I sold it at the time of the great motor-car shortage I obtained £150. A neighbour, who was later killed at Salerno, taught me to drive. This old Ford was the first of a long line of cars that I have driven, but I must say I find tinkering about with them very disagreeable and I have always avoided doing anything with a car except drive it, unless absolutely essential. Within a week or two of purchase of my first car I ran the 'big-end' and after a great deal of trouble getting a replacement I took the whole engine to pieces and put it together again, but I would have to be driven very hard to attempt anything like that again.

8

Forests and Streams Far Away

I HAVE counted myself fortunate in the foreign travel that has come my way. I went to Paris with a fellow student when I was eighteen. We stayed in the Rue de Rivoli, the far end, and we spent a great proportion of our time finding the cheapest eating places, but a day was spent collecting insects at Meaux-sur-Marne. The following year I spent my holiday walking and camping along the Normandy coast and got as far as Rouen. A few weeks before I met Emma I spent a holiday cycling around Holland and attended the International Camping Club Rally at Saxenheim in the lovely and most un-Dutch-like country of the Veluwe. The total cost of this holiday, including the £1 return passage-with-cycle from Hull to Rotterdam, was £8. With Emma I went to Norway twice, once walking and once cycling on the tandem. On the walking trip we crossed the ice fields and glaciers of the Jostedalsbrae and had to hire two guides, and as we had not sufficient money they agreed to be paid after we returned home. We very much admired their trust in us. On the tandem we cycled from Gothenburg in Sweden to Bergen in Norway, and it was some ride, as the altitude at one point was such that the road passed through a lane of ice. It rained hard every day.

Another holiday spent with Emma was a walking tour through the Holsteinischer Schweiz between Hamburg and Kiel, and we visited Emma's grandmother who lived in the latter town. The old lady paid for an aeroplane journey to our next stop, Holtenau, at the mouth of the Keiler Bucht, and this was our first experience of flying.

In 1953 at a meeting of the Freshwater Biological Association, in the lecture room of the Zoological Society, a speaker had called attention to the vast amount of purely descriptive work on fresh-water animals, including insects, which was required to be done in order to attain understanding of the biology of the fresh water, and thus of its fisheries, of Lake Victoria in East Africa. The speaker was R. S. A. Beauchamp, Director of the Laboratory at Jinja, Uganda, of the East African Fisheries Research Organization. This was the time of the early Comet flights to East Africa which had cut down the flight time considerably, compared with the times taken by conventional aircraft. I got up on to my feet at question time and asked, in the days of rapid air travel, when Uganda was only a few hours away, whether it was not possible for the immature stages of aquatic tropical insects, such as the various species of caddis flies, to be collected in their tropical habitat, correctly packed and flown to London, there to be collected and within an hour or so of arrival be swimming about happily in heated aquaria, where their habits and metamorphoses could best be studied, in temperate laboratories. At the end of the meeting Beauchamp came round to me and said how interested he was in my suggestion and would I like to try this out myself if he could find the travel money. Of course, my answer was yes, and within a few months I had an invitation to work at his laboratory for six weeks. I stayed in the bungalow of a young entomologist on the staff of the Fisheries Organization.

My project was to collect larvae of as many different species of caddis fly as possible, bring them back to the laboratory at Jinja and keep them in aerated aquaria until shortly before I left for England, then I would pack them in water weed (they die rapidly in water when transported) and get them into my aquaria at home without losing any time. I would then endeavour to rear them through to the adult stage in order to get the species identified.

I had great difficulty in finding larvae amongst the profuse water weeds and lilies at the edge of Lake Victoria near the laboratory. This was a wonderful site for a laboratory, set on lawns that rolled down to the lake's edge. To the north, the lake narrowed into a channel that finally spilled over a line of rocks to form the Nile. This was the Ripon Falls, a sight of great beauty. One could clamber along the line of constricting rocks to see the Speke Stone commemorating the discovery of the Nile in 1861. Where the water of the Nyanza gathered momentum before swirling over the rocks I found the larvae of two species of caddis in great quantity. In fact, at the edge of the water the dead bodies of the adult flies of these two species were heaped several inches high and their decomposing bodies gave off a nauseating odour which was characteristic, the nearest in quality and quantity being that given off by dead crocodile.

Another collecting place where I had some success was the Nyenga Forest stream at the edge of the Mabira Forest. A path through the forest crossed a stream and the vegetation was dense with creepers and wild date. It was there I first collected adult caddis of the interesting *Anisocentropus usambarensis* species and also its larvae that make cases out of pieces cut from submerged leaves. This and the two species mentioned above I was able, subsequently, to rear through to their final metamorphosis.

The Sezibwa Falls was another delectable spot. A broad pool collected at the base of the falls with a tree-trunk acting as a bridge across the centre. The lower half of the pool disappeared into dense forest, but peering into the water I noticed that the large flat leaves had pieces cut out of them the size of the *Anisocentropus* case, and, sure enough, I found a large number of larvae. I spent a few days at Kampala, where an entomologist friend, Richard, was a lecturer in zoology. He took me collecting to the Namogongo River, the precise spot where Bishop Hannington was murdered. Here we collected in a rather different way. We took a couple of milk churns with us in the car and instead of inspecting carefully the contents of our nets after we had dragged them through the water, the idea was to make as many sweeps with the nets as possible and quickly empty them into the churns. Then we were to get back to the lab and aerated water as quickly as possible and go through all the weed and debris under the better lab conditions. This worked out very well and I found three species of larvae, all of which I reared out later in my aquaria in Surrey.

Beauchamp arranged for me to accompany a small expedition up the river Kagera which flows into Lake Victoria at about the centre of the west coast, not far from Bukoba. Although large, very few Europeans had been along this river, except a few film companies. Our boat was the *Darter*, named after the bird, and our skipper an officer of the Lake Victoria Fisheries Service. Our chief biologist was a South African in the employ of the East African Fisheries Research Organization, now Dr. H. Greenwood of the British Museum, Fish Section. I was just Visiting Scientist, E.A.F.R.O., and we had six African crew and staff.

This was a most exciting and memorable trip for me. I was forty-four and had never been out of Europe and here I was on the bows of a boat chugging its way down river scenes of tropical splendour. We made our way through about a thousand hippo and many crocodiles

One hot afternoon in May, 1954, Philip took me collecting to the Mwiri pool. The Land Rover took the steep, stony and lumpy hill road fairly steadily, then suddenly we could see over the tops of the forest trees that clothed the hillside. The long line of mountains on the horizon was so intensely blue that the scene was one that could only be Africa. Below us were vast tracts of banana shambas, so intensely green that they could only be in Uganda. There were only a few trees on the plateau, but a small depression in the hard laterite held water except for a few months in the dry season. A couple of hours' collecting yielded some dragonfly larvae for Philip, but caddis-fly larvae were absent. But we soon found that this little pool of only a few thousand gallons was full of turtles. I caught three, and took them back to the laboratory at Jinja, but on account of the weight of my luggage I left the two larger at the Zoology Lab. at Makerere College (Richard wanted them). The smaller reached Surrey safely, and was soon the property of Sari. With the water thermostatically kept at 75°F., and worms and ant eggs, it was soon tame enough to paddle itself almost out of the aquarium in its efforts to feed from the hand.

which were all about us, but it was the birds of the tropical African rivers that excited me. I will always remember the goliath heron, flapping away in front of us, the cormorants and darters in flocks, pelicans and wood ibis and the strange bird-calls as night falls. We had our excitements. The current proved too strong for us on one fishing test and we lost a large net. John went off in our small dinghy to retrieve a shot crocodile and as the current swept the dinghy past the

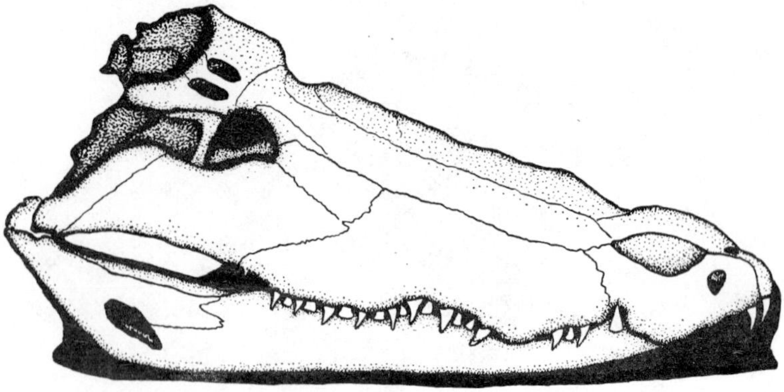

When the University of London appointed me a governor of the new technical college at Ewell in Surrey I felt that I should make some contribution to the zoological collections. In a biology department just starting up it is a difficult problem to amass the great store of specimens required for teaching, so that when I came back from my first journey to East Africa I brought a collection of classical material including lung fish, Protopterus, *and the skull of a large crocodile. When we were sailing back across the great blue expanse of Lake Victoria after the river Kagera expedition, I had voiced the fact that I very much wanted to take back to Ewell the skull of a crocodile. Josepha, the very large Luo who was at the helm, said that he remembered that years ago a hunter had killed many large crocodiles on the shore of one of the rarely visited Sesse Islands, and he thought he could find it. As our journey was to take us only a few miles to the west it was agreed that Josepha could take us there and we would anchor for the night.*

Lake Victoria is the size of Ireland, over twenty-six thousand square miles, and the Sesses are innumerable, but Josepha took us straight to the island he knew and we arrived with only fifteen minutes of daylight still left to us. About a hundred hippos were clustering on the shore ready for their nightly grazing on the lush long grass and they did not like having to take to the water again which they had to when we disturbed their peace. But there on the shore, half on the pebbles, half in the fringing grass, were a number of huge white crocodile skeletons. It was as if I had stumbled into some geological era millions of years past.

Darter he threw the rope fastened to the crocodile instead of the painter for me to make fast. He drifted helplessly by whilst I pulled the crocodile out of the dinghy!

A subsidiary project which I hoped to carry out during my six weeks in East Africa was to collect all the adult caddis flies that came my way. After dinner every night, wherever I happened to be, I examined all the insects attracted by the lights, and collected all the caddis flies. This was sometimes quite a task, especially on board the *Darter* moored at the mouth of the Kagera River. But D. E. Kimmins at the British Museum, to whom I gave the collection, was able to describe a large number of species new to science, and, together with a collection made subsequently by Dr. Philip Corbet, he was able to publish a paper on the caddis flies of East Africa.

I went to Africa a second time in 1956. This was as a result of a successful application for a Royal Society Nuffield Foundation travelling bursary. Using the Jinja laboratory as a base, I wished to extend my collection and rearing of caddis larvae as far as the Kenya highlands to the east and the northern tip of Lake Tanganyika, in the Congo, to the west. In Kenya I went to stay with an old friend in the Timboroa Forest Reserve where he was the manager of a timber concern. He had many times camped with me in Wyre Forest. I did the journey by train from Jinja, the station being on the main line from Kampala to Nairobi and Mombasa. The air at Timboroa station at 9,000 feet above sea-level was keen and cold in the early morning and I was taken a thousand feet lower to his home in the midst of the forest, which consisted of African pencil cedar and two species of *Podocarpus* which are allied to our yew trees but very unlike them in appearance. I spent two days collecting in a small stream and found a number of caddis larvae, three of which I subsequently reared through to the adult stage. D. E. Kimmins honoured me by naming one of them, which turned out to be a new species, after me. This was *Triaenodes hickini*, which was very like the species *Triaenodes bicolor* found in many of our English ponds. (Those that survive and have not been filled in by the local authority.) Kimmins had previously named another caddis after me, *Athripsodes hickini*, which I had collected around the deck lamp of the *Darter* when we were moored at the mouth of the Kagera River. At Timboroa I kept all my living specimens in a bucket (there was ice on it in the mornings, although the Equator was only a mile or two away), but an hour before I was due to leave I packed them into two vacuum flasks wedged tight with water weed. It was a fourteen-hour journey back to Jinja and aerated water, and it was getting hotter all the time. Several times I had the specimens out of the flasks and put them in the wash basin and let the not very cold water trickle over them. I got them back to the laboratory, and fortunately Humphry was at the station to meet me and to rush my insect charges into cool-water aquaria with air bubbles moving the water around. I was relieved, or rather overjoyed, when specimens of three different species started to emerge from their cases to examine their new quarters.

My third visit to Africa was in 1958 and it stands out as the most enjoyable journey, the happiest six weeks I have ever spent. This was our family safari, the holiday of a lifetime. I long cherished the hope that one day, before all the game animals had been killed off and Africa had been cemented all over, that I would mount my own

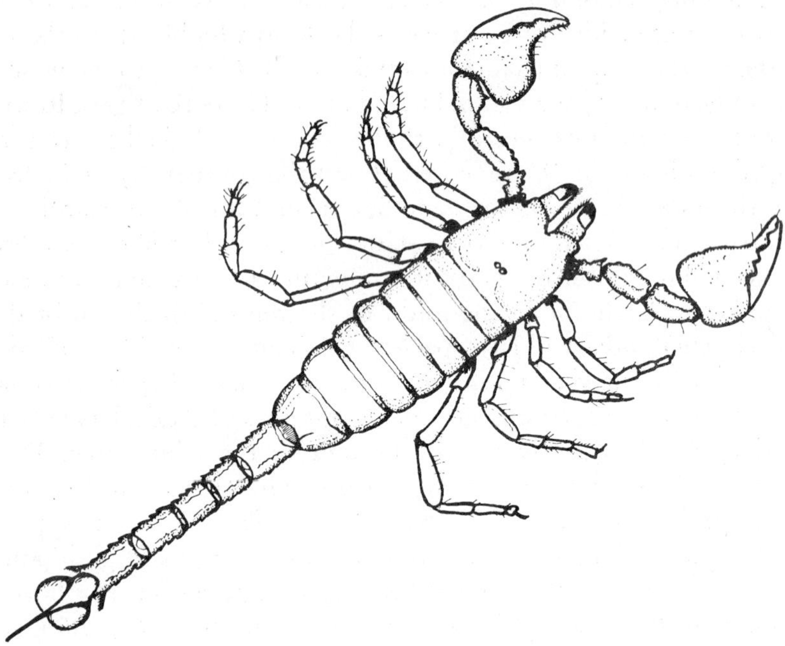

Getting off the rim of the Ngorongoro crater and down on to the Serengeti plain was an arduous task. At first, as the track sloped gently, we wound around the grassy contours, and then we passed through a game-trodden gap. At once the track plunged downwards across the bare shelving rock and it was very dusty, as we could not push the Land Rover along fast enough to keep out of the clouds of dust we ourselves were creating. We had now left the grass-green slopes and we found ourselves in a scorched grey-brown landscape with stunted thorns in a litter of stones and boulders. I suddenly whooped and exclaimed that we must now be in scorpion country: I got down from the car and ran to a largish stone and turned it over with a heave. There, sure enough, was a large scorpion which shuffled slowly backwards as it suddenly found itself revealed to the hot sun, and the underneath of every one of the twenty or so stones that I upended was similarly inhabited.

We first saw Gertie and her calf in the later afternoon, a couple of miles from Ol Tukai, on the dried-up, dusty bed of Lake Amboseli near the southern boundary of Kenya.

We wound the Land Rover in and out of the car-high thickets, with Komati, a game ranger, up in front with us. (What a delight it was to be in the company of a thoroughly trained man.) We had already enjoyed the thrill of seeing a herd of the rare gerenuk giraffe-necked antelope before reaching Ol Tukai, but now the afternoon was drawing on and the intense heat of the day was abating. Then suddenly, without warning, we found ourselves in a glade between the thickets with two rhinoceroses. The large female had gigantic horns and she faced us, raising quite a dust as she scraped the sand with her back feet. We were thirty yards away, and keeping the engine running we seized our cameras. After about a minute she became uneasy and started to move off, and we thought it best to continue our journey.

In another glade we came across fourteen lions, and afterwards photographed another rhino and twenty elephants.

Next morning we saw Gertie and her calf again out on the open lake bed, and she had a large tuft of dried grass stuck on her horn. We learned later that Gertie had a world-record horn. No one had an opportunity of putting a tape-measure along it, but it was estimated to be the longest horn on any rhino living or dead, and to be about sixty inches long.

safari, taking Emma and my two daughters Verney and Sari around East Africa to see the peoples, animals, plants, geographical features and biological research stations.

Early in 1957 it looked as though the time might be imminent when an investment could be capitalized and 1958 would see Verney at her last term at Cheltenham before going to university. She would be eighteen and Sari would be sixteen. I was in correspondence with many people whilst I did the planning, finally buying a 107-inch wheelbase Safari-type Land Rover and shipping it to Entebbe in Uganda. The day after the girls came home from school we went off to London Airport, but a mishap occurred to the plane as it was taking off and

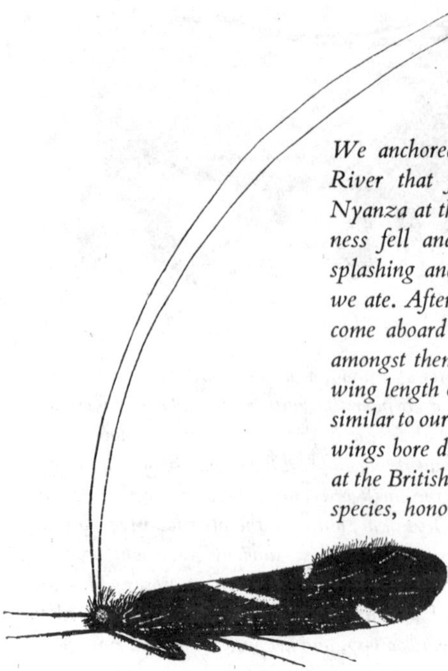

We anchored the Darter *at the mouth of the Kagera River that flows into the west side of the Victoria Nyanza at the Uganda-Tanganyika border. When darkness fell and the hippos had gone ashore with much splashing and grunting, the pressure lamps were lit and we ate. Afterwards I collected all the caddis flies that had come aboard and had settled around the lights. I found amongst them a number of a small black species, with a wing length of only about a quarter of an inch, somewhat similar to our own British longhorned* Athripsodes, *but the wings bore distinctive little white marks. D. E. Kimmins at the British Museum subsequently described this as a new species, honouring me by calling it* Athripsodes hickini.

One of the most exciting finds in insect life that it was my good fortune to make was the rediscovery of Agriotypus armatus *in Britain. This is an archaic parasitic hymenopteron that selects as its host a caddis larva in the sub-family* GÖERINAE. *It is the only known parasite of any of the caddis-fly larvae which is in striking contrast to the state of affairs existing in the* LEPIDOPTERA, *an enormous number of which are parasitized by hymenopterons. The extraordinary thing about* Agriotypus *is that a long twisted strap-like organ is secreted, known as the 'plastron', which serves some respiratory function. I found this insect in a swiftly running Cotswold stream at Miserdon in Gloucestershire. The* GÖERINAE *construct their cases of large sand grains and small stone chippings with a row of small stones placed down each side; they are quite unmistakable, and the long thin brown plastron indicates also quite unmistakably which of them are parasitized. This insect throws light on the probable evolution of the* LEPIDOPTERA. *The most recent classification places the* TRICHOPTERA, *the caddis flies, after the* LEPIDOPTERA, *the moths and butterflies. What nonsense! The very existence of this* primitive *parasitic hymenopteron parasitizing the caddis flies shows beyond all doubt that the caddis flies preceded the* LEPIDOPTERA.

we slid to a halt at the end of the runway. We finally got away twenty-four hours later and at Entebbe we threw our red safari bags into the Land Rover, which was waiting for us, and we were off.

We only had to do twenty miles the first night, as we were to stay at Makerere College, Kampala. The tropical night had enveloped us but a friend, Richard, said he would be waiting at the side of the road to guide us to Makerere, 'Where, over the brow of a hill the lights of Kampala are first seen.' We were very tired but excited as we started off on our long awaited safari which was to take us 4,000 miles and through five countries. At last we breasted the long hill and there were the lights of Kampala below us, and there at the side of the road were the rear lights of Richard's car. Marion, his wife, had prepared a delicious meal: avocado pear and Kenya ham.

From Kampala we set off along the road through the Mabira Forest to Jinja, but we called in at the Sezibwa Falls on the way. Then we drove into Jinja over the newly completed Owen Falls Dam, but the Ripon Falls had now disappeared and the Speke Stone with them, and the Nile now starts its journey to the Mediterranean through turbines and its first appearance is like water coming out of a large-sized tap. After two nights at Jinja we were off to Tororo for lunch with our friends from Timboroa who now lived in this town by the gigantic volcanic plug. As we drove through the forest I have never observed in any other place so many butterflies. Huge swallowtails and a large black and green species hit the windscreen two or three at a time. We were twenty minutes late for lunch and in the evening Ron and Nan and their daughters put on a wonderful party for us.

We got off early next day (in spite of thick heads) and eventually got to Nakuru, where we were to stay the night and visit the Pyrethrum Board's laboratory. I had been warned that we should have a little trouble with the carburettor of the Land Rover as we started to gain altitude in Kenya, but the climb is so gradual that when the engine lost power I could hardly believe that it was strangling for lack of air. At Eldoret the mechanic had everything organized and within minutes we were on our way. We saw our first giraffes at the side of the road a few miles to the north of Eldoret. The next stretch was to Nairobi, where the car had its first service. The following day we spent on a coffee plantation managed by the husband of a former secretary of mine. We went to Nairobi Game Park, an extraordinary sight to see zebra and hartebeest at the edge of the aerodrome, but there were far too many cars near the lions to make this a great pleasure.

Our next drive was to Lake Amboseli and we were to stay in the Ol Tukai game lodge. We stayed there three days as official guests (which we thought was charming) and whilst there had the services of a fine game scout, Komati. The arrangements of this lodge were centred on the habits of wild animals and not the dinner-gong. We had to provide our own food and we had stocked up at Sultan Hamud with a number of loaves. However, when we came to eat some a few hours later they had shrunk to only half their volume and were as solid as rock, but we ate every morsel of them during the following days by soaking them in soup.

Approaching Amboseli, we had seen ostriches on the dry desert-like plain which hardly showed a track for us to follow, and when we came to an area of park-like bushes and trees we saw gerenuk, the giraffe-necked gazelle. This is a bush-browsing species which stands on its hind legs steadying itself with its fore legs to get at the leaves. We arrived at Ol Tukai in a vast cloud of dust at 3 p.m. and an hour later, having unloaded, we were out with Komati. During the next two hours we found ourselves in the centre of a pride of seven lions, had been charged by a rhino, seen Gertie the rhino with the world-record horn, standing out like a great scythe on her nose, and seen a herd of elephant making their leisurely way through the trees stopping now and then to pull a branch down with their trunks. In addition we had also seen zebra, wildebeest, impala and kongoni, one of the hartebeests. The next day we explored the area around the dried-up lake and repeated the feasts of animal observation we had enjoyed the previous evening.

Komati was a remarkable man, thin and wiry, and he wore the uniform of the Kenya National Parks game staff, immaculately laundered khaki shirt and shorts with a green képi. He was very competent and knew the animals he sought out and showed us. He sat on the front seat between Emma and myself and as we approached slowly making a circle he would make expressive gestures with his right hand, a little to the left, a little to the right, and point to the controls which he knew as well as I, four-wheel drive, low gear ratio. Stop! switch off, and there in front of us a herd of elephant would be standing under the trees. Komati knew with great precision the area surrounding an animal to approach within which would be considered a menace by the animal. He knew which animals had a docile disposition, *habana kali*, and which were quarrelsome or truculent, *kali sana*. His official rate of pay was four shillings per day, but when we left, Emma gave him her hunting knife with the handle carved in the shape of a deer's foot.

The last evening at Amboseli, spurred on by Emma and the girls I tried a feat of rough motoring. This was the ascent of Observation Hill. We had already stood by in the distance watching a large American safari car attempt the steep rocky slope and fail, so I was persuaded to try. It was rather foolhardy because if I had broken some vital part of the car transmission system it might have held us up for many days. However, I tried and went up the first time, although I felt I had given the wagon a flogging. What it did do was to give me confidence in the Land Rover over really rough stuff and I think this was the start of real affection which grew between us all for the vehicle, so that when inevitably we had to part with it a few weeks and several thousand miles later (when it was no longer a new car), we felt a grievous loss.

We then went on to Moshi in Tanganyika, going past Masai villages which were like a number of beehives with a high hedge of thorns surrounding them. When we crossed into Tanganyika appropriately the first animals we saw were a herd of giraffe, the national emblem of the country. We spent one day at Moshi, making a foray going eastwards into the Tsavo National Park and thus back into Kenya again. We did not see a single animal, although we collected one of our most generous collections of red dust; we looked like Red Indians, and it was very convenient that we found a garage with facilities for de-dusting the vehicle; no doubt it was common practice. Retracing our steps to some extent, we went back to Arusha and called on the Game Department at Tengeru and the Colonial Pesticides laboratory, where I was surprised to learn that some crops were grown and sold in order to pay for the research on pests. This seemed a novel way of eking out the inadequate Treasury grants, even if the energy and time of the research workers could have been better spent. They had specimen plots of perfectly grown plants, cotton for instance, and it was explained to me that it was not of much use showing how to control crop pests if the local farmers did not know how to grow the crop correctly in the first place.

At Arusha we had to lay in supplies for about a week, as we should not be near a village for some time, then we went on to the game lodge on the rim of the Ngorongoro Crater, but on the way we visited Lake

Manyara. This was an unforgettable sight, the brilliant blue of the water and the dazzling white of thousands of pelicans and lesser flamingoes. The stench of the mud at the edge of the lake is also remembered. A herd of buffalo grazed amongst the rushes a couple of hundred yards away, whilst troops of wildebeest cantered up and down the park-like lawns. We went on to the crater where we stayed three days and we hired Idi, a game scout, to accompany us. Idi was a first-class guide, his speciality obviously was in finding lions. He was constantly scanning the horizon in search of the tips of their large saucer-like ears on top of the grass. When we were with Idi, lions seemed to be around us all the time. Twice when we were out of the Land Rover a maned lion made his appearance, once when we were having lunch and another time when we were all looking at a line of elephants about 400 yards away. In the next instance a fine black-maned lion got up out of the grass and took one look at us and quickly got himself out of sight.

We had a wonderful time in the crater; game was always in view and at times we drove through several hundreds of zebra and wildebeest that just skipped out of our way to let the Land Rover through. Finally we had to make our way to Seranera in the middle of the Serengeti plains and Idi came with us. The journey westwards around the crater rim and thence down a mountain side was rather an ordeal for us all, but we came out on to the rocky and very dusty plain without mishap and we made a brief call at the Olduvai Gorge where Dr. Leakey has made so many discoveries concerning the evolution of man. Two peculiar incidents occurred mid-morning; although the Serengeti appears very flat, actually it is slightly undulating and when not a bush or tree was in sight, and all we could see except the short dry grass of the plain were hazy blue mountains to the north, we observed that a rhino had passed along the wheel marks on the grass that we were following. Mile after mile we followed his three-toed footprints, which showed very clearly on the fine light-coloured dust. Then suddenly he came into view, he was lying down about 200 yards away but suddenly he got up and faced us. I was all for getting a little nearer and taking photographs, but as the ground had deteriorated, being clumpy grass and sand, so that I could not do more than fifteen miles per hour without shaking us to pieces, I reluctantly went on.

The second sight was at noon when we were ten miles from the Lion Rock, a rocky outcrop in the plains with some trees, and we were making towards it for shade for lunch when a mirage occurred.

It appeared that all in front of us was water and the Lion Rock appeared to be twice as large as it was, apparently reflected on to the misty water below. The lunch spot was idyllic and after encountering three lions we reached Seranera and unloaded into the two bandas assigned for us. At the time of the year we were making our safari there was a great concentration of animals at Seranera and in fact they pay little attention to the group of native-type huts which serve as the accommodation for visitors. In the next two days we saw most of the species of large game animals of East Africa. One sight was of several hundred Thompson's gazelle, all running along swiftly and gracefully together. Cheetahs, leopards, topi, lion and hyena were seen within a few hundred yards of our bandas and at night lions visited our bandas and sniffed under the door.

We were next to traverse the Banagi migration corridor leading north-west towards Lake Victoria and we had a report that large numbers of zebra and wildebeest were coming down the corridor, and so it proved. We had not gone many miles when hundreds of these animals were seen making their way towards us. They were twenty to forty abreast and as we went on the front rank divided and columns reeled off right and left of us. This happened continously for thirty or forty miles, an awe-inspiring experience and the sight must have been like the bison of North America about 1830. We came across three lions that got up from a kill—a young wildebeest—as we approached. They did not go away but just sniffed the air as we photographed them. Later that afternoon a herd of roan antelope forty or so strong crossed just in front of us, the first we had seen. We were amazed that they were so large. There were some magnificent bulls amongst them with long back-sweeping horns. I couldn't get the Leica out quickly enough for photographs, so I turned off after them. However, the going became rougher and my passengers started to complain of being thrown about, so I got out and ran after them. Unfortunately, getting to photographable distance was impossible and I trudged back out of breath. I read afterwards that it was dangerous to follow roan when you have frightened them, as the master bull will turn and come for you with his head down if he thinks his cows are in danger. We also saw a herd of about twenty eland before we came on to the road which follows the water edge of Lake Victoria; we had only seen small groups of two or three before. We then drove without incident into Mwanza, a port on the Nyanza.

Our next destination was Shinyanga, the tsetse-fly research station, where many well-known entomologists had worked. The road ran due south through dry shrubby country, almost desert-like in many places, and at last we arrived there. The research station is situated in an old German boma or fort, stone below and corrugated iron on top, a relic of former days when the country was a German possession. Research workers and their families lived in bungalows scattered around. The Director, Dr. Glasgow, and Mrs. Glasgow, made us very welcome and a party had been arranged for us that night. Apparently it was a rare event in Shinyanga for a visitor to drop in *en route* for the Congo with three ladies, two unmarried. I spent the next day in the research station but before dusk we went to the hill near the boma. This was of indescribable beauty. A mass of great boulders covered with bougainvillaea, in various shades from deep maroon to light orange, frangipani and many rare plants besides. In years gone by it had been the burial place of tribal chiefs and was a holy place. In recent years entomologists who had died or been killed at or near Shinyanga had been buried there. Swynnerton, a previous director, killed in an aeroplane accident, is commemorated by a plaque fastened to a boulder. His great work, the *Tsetse flies of East Africa, a first study of their control*, published as Volume 84 of the *Transactions of the Royal Entomological Society* in 1936, runs to 579 pages, 22 plates and 7 maps, and is a monument to his great work for Africa. Those who followed him, some of whom I number amongst my personal friends, have followed in the great traditions.

As we came off the hill we sat for a long time quietly on a boulder and when I think now of Africa I think of Shinyanga; I think of the African we could hear nearly two miles away walking to his home, a few sticks in the ground and some dried grass over it. I think of Shinyanga, where we had surprised two Africans filling their water skins from water contained in a concavity in a huge rounded boulder. It was green and it stank, fouled by baboons, with their excreta floating in the water. We followed them with our binoculars for an hour afterwards, picking their way very carefully over the hot sharp rocks in their bare feet with their precious cargo of stinky fouled water.

We never found anyone who had driven into the Congo by the route we wanted to take, but we had written to the district officer of Kahama and the district commissioner of Birharamulo, and a hotel at Kitega in Ruanda-Urundi. From Shinyanga we drove through the forest of baobab trees. It was a strange drive in a lost world and we arrived at Kahama in mid-afternoon. The Union Jack was flying over the administration buildings where the district officer was holding court. He adjourned the proceedings in order to direct us to the rest house. This was the only place on our safari where we had to take the seats out of the Land Rover in order to furnish our sleeping quarters, but we were able to get baths. The land between Kahama and Birharamulo was mainly uninhabited, as it was a sleepy-sickness area, but we occasionally passed small bands of Africans with various equipment in their hands. They all looked extremely guilty, but we had no idea of what they were up to. Then we drove up to the boma at Birharamulo and met the Hildersleys, who made us very welcome and set us on our way to Kitega with a large demijohn of boiled water and large quantities of fresh fruit.

During our family safari there were a few disagreeable interludes; I suppose it would have been a miracle if there were not. The first two followed closely on each other and were examples of the unhelpful attitude adopted by Belgian officials towards foreigners, which has often been commented upon in books and articles. When we arrived at the border post between Tanganyika and Ruanda-Urundi an African official took our passports and then proceeded to copy almost everything in them into a book. At last he decided that we must accompany him to the Immigration Office at the town about twenty miles away. We were to follow his car but it would not go until we all got out and pushed it down a hill. At the office we were told to wait, as the official was busy; they appeared to be counting money. In spite of my friendly, frequent remonstrations they ignored us and for the three middle hours of the day we had to sit sizzling in the heat in our parked car. At the end of that time I stormed in, and, making sure that none of the ladies of my safari were in earshot, demanded to be taken to the chief administrator of the district or the chief magistrate. I told them what I thought of their manners, courtesy, administrative ability and competence in a few but well-chosen words. We were on our way in five minutes.

The next day we had to get to Usumbura by noon in order to cash travellers' cheques into local currency, as it was Saturday, when the bank would be closing for the day. We thought it would be a fairly easy drive, eighty-four miles or so, but we had not reckoned with the contours. The road ran up and down and in and out along a section of the rift valley escarpment. At one point we went so high that we went past a clump of giant lobelias, similar to those found on Ruwenzori. As time went on I was wrestling with the fully loaded Land Rover round the hairpin bends with fearful chasms below us and I got more and more worried that we would not make the bank in time. But with a final burst of speed down the last series of hills I pulled up outside the bank just before ten minutes to twelve and the girls got out and went over to a café on the other side of the road. 'Get me an iced beer,' I said as I went through the swing doors of the bank. I got out my travellers' cheques and the manager came over to me. He regretted very much that the bank was now closing and would I mind coming in on Monday and he would be at my service. I pointed out that it required more than ten minutes to noon, but he countered by saying that the balances had been made and the staff were already getting ready to leave. This time it did not take me very long to get worked up. I momentarily thought of the long, hard worrying drive and my family waiting expectantly on the other side of the road with their drinks. It was still only five minutes to twelve, so I threw away any respect I had for the dignity of the bank and made myself look as big and rough as my sweaty and dirty sixteen stone would allow and I gave a series of resounding slaps on the desk that stopped everybody in their tracks (Was this a hold-up? they thought) and I shouted, 'Well, you can bloody well unbalance your bloody balances,' to the shaking manager, who did not know whether to push the alarm bell or not. Indecision came over him. I should be holding a gun, but I was thrusting a £100 travellers' cheque into his face. He gulped, licked his lips and then went to a safe and counted out to me the required amount of Congo belge currency.

As I sipped my beer, Emma, sensing something wrong, said, 'What happened?'

'Oh, I just had a few words with the manager, he wanted to close,' I said.

We visited the Freshwater Biological Research Station at Uvira and occupied the suite on the front of the first floor of this magnificent building. We journeyed northwards from Uvira and our destination was Kisenyi on the northern end of Lake Kivu. There was a magnificent road joining the northern end of Lake Tanganyika with Bukavu at the southern end of the lake but the road around it was quite something. The lake occupies the floor of the rift valley and the western shore was precipitous. For a mile or two we would be driving along the shore then we would wind our way up and up until we were several thousand feet up. When we arrived at the northern end of the lake it was dark and a thunderstorm broke just as we were unloading at the hotel.

It had been our intention to stay a few days in the Parc Nationale Albert, but we did not feel altogether happy in the Belgian Congo so that at this point we decided to change our plans and get back to Uganda via Kisoro and Kabale. This was a wonderful drive, the road taking us to a considerable altitude and for miles we went through bamboo, catching sight of the volcanoes Muberga and Muvira. We knew this was the gorilla country. At the Uganda frontier we met Mr. Baumgartel and had lunch in his hotel. He was the expert on seeing gorillas and wanted us to stay a few days with him so that we could be taken up the mountain where they are to be found on the Uganda side. After a counsel we decided to move on as, although we had unspent Congo money, we had not much East African. On our way to Kabale we climbed a high mountain road, the Kanaba Gap, which took us up to well over 8,000 feet, and as we came through the top of the pass we had a breathtaking view of Lake Bunyoni, 2,000 feet below us and encircled with mountains. We spent the next afternoon on its shores, although it rained most of the day. At Kabale we stayed at the White Horse Inn and there met a staff officer of the Uganda Game and Fisheries Department. He was on holiday there with his wife, and when they learnt the details of our safari, the remaining part being to visit the Queen Elizabeth Park and stay at Mweya Lodge, we were urged on to Mweya and then to visit Paraa Lodge on the Nile near the Murchison Falls, as we should there be able to photograph crocodile which we had so far been unable to accomplish. He was sure they would be able to accommodate us, so instead of resting we girded our loins and, armed with East African currency, on we went to Mweya by a new route which was drawn on a map for us.

By mid-morning we were in game country and at lunch we were within the confines of the park and seeing herds of animals grazing in the meadow-like pastures. During the afternoon we approached the Kazinga Channel linking Lake Edward and Lake George and we saw large numbers of buffalo. They grazed on both sides of the road like farmyard cattle.

We expected to cross the channel by ferry, but in fact it is spanned by a section of the bridge that once crossed the Thames near Blackfriars during the last war. Shortly after crossing the bridge, however, we ran into a large herd of elephant and I ran out of petrol at the same time. Although feeling a little shaky, I took an emergency can of petrol and put it in the tank within sight of the great beasts. At Mweya we drove around the environs of the camp in the morning and in the afternoon we went on a launch up the channel to get hippo and elephant photographs. Hippo were doing so well that many of them were being shot, as the numbers were so large they were beginning to erode the banks and drive other animals away because there was no grazing. We spent a day exploring the explosion crater area, fifty or so extinct volcanic craters just separated from each other by a knife-edge track, but only on two or three was it possible to drive the Land Rover. We also inspected the very large herd of buffalo that grazed on a wide flat plain. To the west we could see the Ruwenzori Mountains.

Next day we carried on northwards past Lake George to Fort Portal with the foothills of the Mountains of the Moon on our left and never out of sight. From Masindi onwards we went through an area of tall elephant grass. There had been a little rain in the district the previous day and where the track ran into a little valley a truck was stuck and already holding up two other cars besides ours. After a certain amount of talking, the girls and I cut a lot of branches and laid them on the mud about twenty feet further along the track from the truck. Then I drove round the cars and set my wheels on the branches and tied our steel wire (which we carried so that *we* could be towed out of mud) and then towed the truck out as though this was the sort of thing we did every day. Later in the afternoon we crossed the Nile just before it enters Lake Albert and settled in two tents, the only accommodation available for us at Paraa.

Whilst we waited for the pontoon to come across the Nile to fetch us it was very hot. Indeed, about a hundred yards away a buffalo was having a cooling bath, only his nostrils and horns were visible out of the mud he had churned up, whilst on the other side of the river an elephant sloshed about amongst the reeds, and standing motionless was the only shoebill stork we had seen.

The two days we spent at Paraa Lodge were full of activity and excitement. We drove along the northern end of Lake Albert and saw where the Nile left this lake for its long northwards journey through the Sudan and Egypt to the Mediterranean. We saw two animal species in fair quantity that we had not previously encountered: oribi and Uganda Cob. Both animals were scattered along the lakeside, resting amongst the clumps of reeds. They got up with surprise, gave us disdainful looks and then bounded off. Always this happened so quickly that although we approached fairly closely to them our photographs only show a series of backsides. Elephant were very plentiful and our guide on one occasion showed us how to stampede a herd by making a few handclaps and then shouting, 'Coo, Coo, Coo,' not very loudly. The herd of about thirty individuals gathered itself together and went away at great speed. We drove to the north bank of the river to the Murchison Falls where the Nile hurls itself through a nineteen-foot slit in the rocks to the troubled waters below. We went by launch also to the falls along the river and we were able to take all the crocodile pictures we wanted. The birds were also a wonderful sight. This great tropical river was lined with dense high forest. Fallen trees lay half in the water on which perched the goliath heron and crocodiles were motionless, like scaly logs in the water or lay on sandbanks with their mouths open. I can now only see this in retrospect but none the less vivid for that. One night hyenas came into our tents and took away a large straw bag containing tinned food and scattered it around. On our last morning, just as I was shaving, Sari came to me and whispered, 'Dad, there's elephants all around us,' and sure enough their great hulks were moving through the surrounding bushes with seesaw-like motion. I managed to photograph Sari with seven elephants around her. (No doubt a picture of me lathered and taking photographs would have been more amusing.)

Then came the time for us to get ferried back across the Nile and to make our way to Kampala. We were held up several times by groups of elephant on the track, but then we skimmed along, landing back at Kampala at teatime. A mad rush started, all the things we wanted to take back home, including skulls of rhinoceros and giraffe, a huge tortoise, etc., etc., had to be packed. Then we waved goodbye at Entebbe Airport and we were on our way home: Verney to university and Sari for her last two years at Cheltenham.

When we disembarked I said to my family that this was the longest time we had all spent together and we were still on speaking terms with each other!

9

Professional Entomologist

LIKE very many other men working in scientific laboratories during the war I was carrying out programmes of work in disciplines to which I was not naturally inclined. It is true, however, that I had been brought up, so to speak, as a laboratory assistant in chemical and physical laboratories, and also I had to study many engineering techniques during my period of employment on development work on aircraft accessories. During this time I had become qualified as an entomologist. Indeed, soon after I had graduated I wrote after a number of appointments in biological laboratories, but failed to get accepted.

At the beginning of 1944, however, I thought that the time was perhaps opportune to start looking around again and in April that year I was appointed as entomologist to an insecticide-manufacturing concern in London known as Rentokil Ltd. The Central Register of Scientific Manpower, however, would not give their consent to my taking up the appointment and suggested to me that I negotiated an agreement with the firm on the lines of my appointment being taken up at some future time when the consent of the C.R.S.M. would be given. The managing director of Rentokil Ltd. was Miss Elizabeth Eades and she readily agreed to this procedure, so that it was not until December 1944 that I was released from Dunlop's and allowed to take up my first and only appointment in the insecticide industry.

I do not propose in this chapter to do more than outline the biological work which I carried out in the course of my professional duties with Rentokil Ltd., and the companies which followed, with similar if not identical names. The way in which the company managed its affairs, grew and eventually became a byword, I intend to write about as *The Woodworm Story* at another time, and in another place.

The first step which Miss Eades took on my agreeing to join the company was to send me examples of all the products which were made, and specimens of all the leaflets, advertising and other printed material. I found that many products were produced, but all were to be used against insect pests indoors. There were products both in liquid and in powder form for moths and other textile pests, cockroaches, flies, silverfish, bedbugs, fleas, the parasitic lice of dogs, cats and numerous other insect pests, not forgetting to include woodworm. However, my first self-appointed task was to read all the scientific, especially entomological, literature that I could procure concerning all these pests. Especially I wanted to find out how important were these different insects as destroyers and foulers of food, wood, wool, health and peace of mind. I found this an extremely difficult task, as no one had apparently made accurate surveys of the incidence of these pests in our homes. Surprisingly enough, little had been written concerning the biology, the length of the life cycle and all the myriad details about the natural history of these insects. Some of this has, of course, been remedied since, but that was the situation as I found it at the end of 1944.

It seemed to me that I had taken on a Herculean task finding out what was known concerning such a wide range of insects so that I could speak with some knowledge of the particular insect problem and, further, supply a particular insecticide. It appeared to me also that, to resolve the particular insect problem, the specific habits of the insect must be known before the correct choice of insecticidal ingredients could be made. After only a month or two's employment by the company I had rather a shock: this was to find that the turnover enjoyed by the company was extremely small. Learning that the previous year the total turnover was only £18,000, the conclusion was reached that neither I nor the company could afford my services entirely as an entomologist, and indeed it was not very long before I was taking a larger and larger share of responsibility for the running of our little factory. The company had been bombed out of its two previous premises and now occupied two shops, with houses at the rear, in Stockwell, near Brixton, but I had a large airy room in which to conduct my compounding and entomological experiments.

Manufacturing problems at that time were many and varied. Shortages of almost everything meant a constant struggle to maintain supplies and quotas of essential chemicals and solvents had to be negotiated with skilled bureaucrats. I thought that it was essential to play a decisive part in increasing turnover, and in this I had the support and, indeed, the active encouragement of my managing director, Miss Eades. I wrote leaflets and articles in the trade Press and took part in our meetings with advertising agents, and later wrote the 'copy' and made drawings for illustrating them. Quite early on in my survey of insect pests indoors I had become aware of the very much more general occurrence of wood-boring insects than one would be led to suppose from accounts in entomological papers. This was first of all a rather vague feeling or 'hunch', but later became a much more definite opinion, and I began to read all the scientific papers and books on this subject.

Another of the early tasks that I set myself after I took up my appointment with Rentokil Ltd. was to assess the efficiency of the various products by means of biological experiments. I carried out tests, many of them rather rough and ready by the standards of a few years later, with flies, bedbugs (I was able to obtain extraordinary numbers of these from a public health inspector), spider beetles (*Ptinus tectus*), cockroaches, brown house moths (*Borkhausenia pseudospretella*), grain weevils (*Sitophilus granarius*) and the powder-post beetle (*Lyctus brunneus*). The latter I thought at that time to be the most important of the wood-borers because of the constant reference to it in the trade Press, and because it attacked new furniture. Commercial insecticide formulations at the end of the war were largely based on the organic thiocyanates which went under the trade names of the 'Lethanes' and I found them to be very effective. I managed to get hold of a small laboratory sample of D.D.T., the new miracle insecticide of which I had heard a great deal and was in a hurry to try out. I was only able to do a few simple tests with it, but with good fortune I hit upon the best possible formulations for dealing with many household pests. This was in the use of formulated insecticides where D.D.T. was combined with other materials such as pyrethrum powder, or concentrate, and lethane. Pyrethrum, which is made from the flowers of a daisy-like plant, was, in fact, in very short supply and its use had to be extended as far as possible. My tests with *Lyctus*—incidentally the first wood-boring insect with which I experimented—were made with the old-established product Rentokil Timber Fluid, as it was called. I had come into possession of a large drawer-front made of chestnut, which was nearly all of sapwood and peppered with flight holes of *Lyctus* throughout its length. I cut this wood into a dozen small pieces, each about three inches by six, and half of them I kept as controls and no Rentokil Timber Fluid was applied to them. Of the other six, I applied one coat of Rentokil Timber Fluid to three, and the same product to the remainder, except that I had added to it 1 per cent D.D.T. The result was extraordinary. No further flight holes appeared in the pieces of wood to which either Rentokil or Rentokil plus D.D.T. had been applied, but the 'control' pieces were literally reduced to a fine flour except that a very thin veneer of outer surface, and the cellulose polish of one surface, was not consumed. This test convinced me that the old formula of Rentokil was very efficacious, in fact so good was it that the test failed to show me to what extent D.D.T. improved it. The pieces of relatively sound wood on the one hand, and the glass jar containing the fine powder on the other, were shown at exhibitions, demonstrations and lectures for many years, and must have played some part in convincing a large number of people that woodworm *could* be cured.

One of my conclusions after surveying the various household insect-pest problems was that there existed a strange belief that woodworm could *not* be cured. In fact it was very widely held, more especially by the old-fashioned type of cabinet-maker craftsman, that there was only one thing to do with a piece of furniture in which woodworm holes appeared, and that was to burn it. It took a prodigious effort to change this old bit of folk-lore, but change it we did, and this story is told elsewhere. One aspect of this problem, however, I intend to relate here.

I met the editor of the trade journal *The Cabinet Maker*, and he suggested to me that, as part of my project to change the prevalent attitude concerning the burning of woodworm-infested furniture, I should write a series of articles illustrated with line drawings and photographs on the whole subject of wood-boring insects. He said that he would give me a few tips to help present my story to the furniture trade and would publish the articles. I worked hard at the articles, which were all published, and there was a large demand for reprints, several thousand sets being distributed. It was out of this that the idea of a little book was born, entitled *Woodworm, its Biology and Extermination*. Unfortunately, owing to a series of unhappy delays in its production, it did not see the light of day until 1954.

I was beginning to make quite a song about the importance of the furniture beetle, *Anobium punctatum*, as it was by far the most important of our wood-boring beetles, not only in furniture, as its name implies, but in the softwood structural timbers of our homes. The entomology department of the Forest Products Research Laboratory was first set up to deal with powder-post beetles, for the High Wycombe furniture trade. It then went on to deal with death-watch beetle. Research on this insect was of great complexity on account of the long life cycle which could be upwards of ten years when its foodstuff, oak and chestnut, is not sufficiently fungally decayed. After death watch, the Laboratory went on to carry out further work on powder-post beetles, then house longhorn, and a special appointment was made in order to carry out work on the latter insect, of which only about sixty infestations are found annually.

It was my contention that this order of priority had been all wrong and that the little brown beetles, *Anobium punctatum*, were by far the most important of our insect pests of timber. Unfortunately, it was thought in some quarters that this was an advertising stunt and that somehow or other, by subtle advertising, we were persuading people to treat their furniture and floors against an insect that we had almost evolved ourselves. Of course, this was nonsense.

A new industry of great economic importance, wood conservation, the conservation or preservation of our great national investment of timber in buildings, was coming into being.

To return to the book, I was eventually persuaded to publish it myself and to appoint Simpkin-Marshall as distributors, which I did. The day after delivery was made of the first thousand copies of the book they went bankrupt. In spite of the first steps taken on the wrong foot, the little book has now completely sold out and it has gone into almost every country of the world. It had an importance far greater than I could have imagined when I laboriously wrote out my first articles and made the drawings.

I ought to say something about wood-boring insects found indoors in Britain, even though I have written so much these last few years on the subject. But as these insects have played such a large part in my life, for over twenty years, the readers of this account should have some idea of the applied entomology of this subject. Even though a large number of wood-boring beetles have been found from time to time boring out of woodwork in buildings in Britain, the number of species of substantial economic importance is small, no more than five insects coming into this category. By far the most important species is the common furniture beetle, *Anobium punctatum*, and we now know for sure that it accounts for about 80 per cent of all insect damage to timber in buildings. In 1944 and 1945 this important fact was only just dawning upon me and its great importance as a pest of building timber was not recognized. Indeed, it has only recently been possible to state that it is virtually certain that over 50 per cent of all buildings in the United Kingdom support an infestation by this little brown beetle. The year before I took up my appointment with Rentokil Ltd., as an entomologist, the total annual turnover for the sale of Rentokil Timber Fluid, specifically for the control of furniture beetle, was a few shillings over £4,000 and no actual servicing work was carried out by the company. Today the annual turnover in servicing work alone is reckoned in millions.

Of the other species of wood-boring beetles found in buildings, let us now consider the death-watch beetle, *Xestobium rufovillosum*. This species, first of all, is remarkable for the amount of publicity it gets but it is responsible for less than 5 per cent of all the woodworm infestations. This is due to the fact that it is found infesting only old hardwoods, such as oak and chestnut, which have already been attacked by a wood-rotting fungus, either a species which was present in the living tree such as *Fistulina hepatica* or *Stereum sulphureum*, or a species which has invaded the timber since it has been incorporated into a building. It seems certain, however, that in every building where death-watch beetle is found it has been introduced from out of doors with the original timber. Death-watch beetle is a common insect out of doors, where it is usually to be found in unsound timber in large over-mature trees, and it is in large dimensional timbers cut from such trees that death-watch larvae get transported from parkland to church or other large old buildings. It is for the precise reason that such large-dimensioned timbers have not been used in buildings for many years that it can be stated that there have not been any new infestations by this insect for perhaps fifty years, so that we can consider that, with the present scale of control work, it is being exterminated.

Of even less importance than death-watch amongst our domestic wood-boring insects are the beetles known as the powder-post beetles, which belong to the genus *Lyctus* of which only one species, *Lyctus brunneus*, is of any significance. The biology of *Lyctus* beetles is very different from the other species so far considered. In the first place, the female beetle will only lay her eggs in starch-rich sapwood of wide-pored hardwoods. The latter not only include our own native oak and ash, but in addition a large number of tropical timbers. There is a specific mechanical reason why the pores of the wood require to be of a certain minimum size; this is because the long egg-laying tube of the female is extended several millimetres into the pore itself and this is where the long, narrow eggs are laid, either singly or in small batches of two or three. But before laying her eggs she makes sure that the special nutrient material required by the larvae is present. This is the starchy cell contents; the larvae are not able to digest the tougher cellulose of the cell walls. The female beetle, when she flies on to a piece of hardwood timber, usually in the timber yard, runs on to the sapwood edge and tastes the wood where it has been sawn and she is able to detect, presumably by taste, if the required content of starch is present.

Lyctus beetles were much more common in the years immediately following the war than they are today. Surveys of the comparative incidence of this species show that it is still losing ground and it seems very probable that the programme of spraying of susceptible timber stocks with chlorinated hydrocarbon insecticides, initiated by the Forest Products Research Laboratory, has brought this about. Because this is a hardwood insect it is, of course, only confined in buildings to hardwood floors, joinery and furniture, and has, in the past, been troublesome in such cabinet fixtures as shop fittings. However, it never was an insect anything like as common as the furniture beetle, nor was it ever such a potential pest for the biological reasons already explained.

Another wood-boring insect found in buildings is the house longhorn beetle *Hylotrupes bajulus*, but, in strong contrast with its almost worldwide distribution, it is confined in Britain, at least as a breeding insect, to a relatively small area of north-west Surrey. It has been found in a number of other parts of the country, but only in situations where it has been introduced in timber as larvae, which subsequently die out. This species is known to come into Great Britain from European countries in orange boxes and other lower-grade timber. The house longhorn infests only softwoods and generally only the sap, although when this has been consumed to the extent that it is a mass of tightly packed frass, it will turn its attention to the heartwood. It is such a large insect, the larva being a fat legless grub, an inch long, that it only requires a few of them to cause appreciable damage, and when they are gnawing and the house is quiet their jaws can be heard scratching away. It is most usual for a house longhorn infestation to take place in the roof void of a building, a situation where it is hot in summer and cold in winter, and perhaps it is significant that its distribution as a breeding insect in the United Kingdom occurs where the highest summer temperatures occur.

The last species that I wish to deal with in this account are the wood-boring weevils. Two species are involved and their scientific names are *Pentarthrum huttoni* and *Euophryum confine*. They are small blackish insects of typical weevil appearance, that is, there is a long rostrum or snout at the front of the head and the elbowed antennae are borne halfway along it. The legless larvae of both species are found burrowing into wood in which substantial fungal decay has already occurred. This means that wherever they occur, the problem is entirely a wood rot problem in which fungicides should be involved. The first-named species is probably indigenous; it has certainly been with us in the United Kingdom a very long time. On the other hand, the last-named species was first recorded in this country as recently as 1937. Since that time it has spread with great rapidity; not only is it now found throughout Britain but in London and the south-eastern Home Counties it is virtually found wherever a substantial attack by wet-rot fungus, *Coniophora cerebella*, occurs. The adult beetles are incapable of flight and the legless larvae do not wander about, being confined to their tunnels in the rotten wood so that how they have effected such a wide distribution is a mystery.

The great increase in the population of this insect has occurred since I have been concerned with wood-boring insects and the actual method of dispersal of this species of wood-boring beetle and, indeed, of furniture beetle and death watch too, have been of extraordinary interest to me, although in the case of the death-watch beetle I think the problem has been resolved in that it has not dispersed at all.

Immediately after the purchase of the Rentokil Company by the British Ratin Company, I conceived the idea of writing a comprehensive book on the wood-boring insects likely to be found in buildings in Britain and illustrating all species which were likely to turn up in such circumstances. In those families of beetles where a few species only would be found in wood in buildings, whilst the other members of the family would be purely forest insects, or perhaps infesting stored food products, I would describe all of them so that a handbook would be available for all who had to identify such insects and give advice on their control.

I planned out the work carefully in a paper which I called 'my written work', which I circulated to my colleagues, and I thought that it would take me three years. In fact, it took just over four years to write and nearly a year to get through the Press. This is my book, *The Insect Factor in Wood Decay*, which Hutchinson published in January 1963. My plan was to relate all the written papers that I was to produce for the company's pest-control conferences, the British Wood Preserving Association's News Sheet and Annual Convention and articles in the journals *Pest Technology* and *Timber Technology*, and a paper for the Institute of Wood Science, to this single purpose. In this way I monographed one group of insects and subjects one after another, but the illustrations took up a great part of my time. The scraper-board drawing of *Coptops aedificator* took me almost twenty-four hours to complete. Altogether the book ran to a hundred thousand words, excluding the references of which there were many hundreds, and 263 illustrations, the majority of which were indian ink on Bristol board or scraper-board. I thought Hutchinson's made an extraordinarily good book of it, the paper being excellent. If any criticism could be justified it was that the line-drawings were over-reduced. The British Wood Preserving Association published my series of papers on the British longhorn beetles (*Cerambycidae*) where the illustrations had not been reduced to such an extent and the comparison is remarkable. The whole of the writing of the book and all the illustrations were drawn in what could be called my 'leisure time'. This was from about eight o'clock in the evening until just before midnight, Saturdays and Sundays, and holidays. Although most of the work was done at Home Farm, Fetcham, when I was beginning to have my arm twisted to get it finished on a date-line, I would take a few cabinet trays of beetles to the Newalls at Wyre Forest, and draw early and late with only the far-away country noises to interrupt.

When I was finishing the *Insect Factor* I dangled a carrot in front of myself to drive on to the completion of the writing. Emma and I would take a long holiday, which we would have earned, but in fact no such thing occurred! Immediately I finished this major work I set about three smaller books, *The Woodworm Problem*, which was to replace the original little book, *Woodworm—its Biology and Extermination*, a companion volume, *The Dry Rot Problem* and *Household Insect Pests*. These were published in 1964 and 1965.

During this period, 1958 to 1963, it is not surprising that I did practically no work on caddis flies, except that I wrote a short paper on caddis larvae that I collected in East Africa during my tenure of the Nuffield Foundation travelling bursary.

Returning to the time when I learnt that the Central Register would release me so that I could join the Rentokil Company, I started to make enquiries about living in the London area. Later, when the date was agreed when I should take up my appointment, I spent all the time I could afford looking for a suitable house in various districts. I must say I was appalled. Of course, I did not know the various residential districts of London at all, and as I quartered them one by one I was amazed at their general dinginess, noise, overcrowding and lack of local genuine countryside, and, perhaps most of all, houses of any sort were so expensive.

It was soon realized that neither Emma nor I were going to be happy within ten miles of Stockwell where the Rentokil factory and office were situated. I then made the decision to find out the quickest method of getting out of London at night, so that I went to Victoria Station and studied time-tables. I was immediately impressed with the semi-fast trains to Brighton that stopped at East Croydon, Redhill and Haywards Heath before reaching their destination, and the following Sunday I took the first train out of Victoria on this line. I got out at Redhill thirty-five minutes later and immediately walked to the left, Redstone Hill, on to higher ground. This is the main A25 road, but it was quiet, not much traffic and the wartime-restricted bus service still operated so that no buses ran on Sunday mornings. I walked on through the village of Nutfield and open countryside came into view along the high ride which the road follows.

This was how we got to the village of Bletchingley in Surrey where we lived for over ten years. On that Sunday morning I found a large derelict house of considerable age and after making some enquiries about it at the Whyte Hart next door, I arranged for Emma to come down from Birmingham to have a look at it. She was not the least bit interested, but learnt that a house was shortly coming up for sale at the top of the main village street; it was a Victorian house with white pebble-dashed walls, six of the rooms facing south, and containing long halls and landings. It was called 'Plummers'. We bought it, and after I had painted and papered it throughout we moved house.

It was the time our two daughters Verney and Sari were growing up and we were happy there. It was the time also of great physical and mental effort in the part I played in building up the company as well as doing my caddis-larvae descriptions; indeed, the bulk of my published work of this group of insects was carried out at 'Plummers'.

The main A25 road bisected the village from west to east and with the later years the growing volume of traffic made walking in these directions virtually impossible, but north and south was very different. A few yards from the road we could take the children for picnics and collect flowers, fungi and insects. Although there was very little really wild country, there was an abundance of wild life.

Bletchingley is a fairly large village situated at an elevation of 500 feet on what is known as the Nutfield Ridge, a quite distinct but parallel hill to the main mass of the North Downs. Now badgers are very common on the North Downs, but nowhere are they more common than in the Bletchingley district. Because of their strict nocturnal habits they are, of course, seldom seen alive, but traps and fast motor-cars travelling at night take a toll and poor Brock's lifeless form is seen by early risers.

This story, however, although it has the same unhappy ending, tells of a badger which, being old, and unable to get a living on the hill under the castle grounds, crossed the main A25 road and took up his abode in OUR GARDEN. It was some time before we found out what animal had adopted us, but eventually we found that a very old badger, with scarcely a hair on his back, had excavated an inconspicuous hole under a woodshed next to the garage, and in the daytime he slept there. He was some twenty yards from heavy constant traffic and only a couple of yards from our own car, which was driven in and out of the garage every day. For some months he used our garden and the adjacent school lawn for his survey of the local earthworm population. We just had to put up with the constant scratching (Brock always uses his left forefoot for this, as shown in the drawing), and we resigned ourselves to the fact that one of his favourite worm-scratching grounds was where we were trying to get a few rows of young cabbage along.

Every night he came inside the verandah at the back door of 'Plummers' and helped himself to any meaty (or fishy) scraps that were in the chicken food bucket. Quite regularly, half an hour after dark, we would hear the bucket fall over as Brock's heavy body leaned over eagerly to sort the scraps over with his snout.

Then he was seen in the village, first by a girl delivering newspapers, then visits were made to flash a torch on him on the school lawn. But

The old badger that lived in our garden came to the back door and hunted for any scraps which he fancied in the chicken food bucket. He would then hunt for worms on the school lawn next door.

The earth stars (formerly called Geaster, *now called* Gaestrum) *constitute a small well-defined group of fungi within the order* Lycoperdales *and thus are related to the puff-balls. Twelve species have been found in England but all are rare. They differ, amongst other characters, in the number of rays into which the exoperidium splits. The illustration shows the 'arched earthstar' with a four-rayed exoperidium, the tips of the rays still adhering to the cup-shaped inner layer.*

Getting some old oak gate-posts out for replacement the other day I was reminded of the hot July evenings in this Surrey village of Fetcham. Eating quite deeply into the rotting, earthbound end of the post were several large grubs of the stag beetle. Not even the most fervent coleopterist could say that these fat, slowly writhing larvae in the shape of question marks were beautiful; yet, on contemplation, how well-fitted they must be for their way of life. Nearly two feet down in the wet ground, cold and entirely lacking in light, the oxygen tension low and yet with little protection, they feed on the rotting wood for several years before finally flying about in the summer dusk as a great antlered beetle.

In contrast to the large number of beetles whose larvae burrow into and feed on wood the Lepidoptera, *the butterflies and moths, are poorly represented as cellulose feeders. But the wood leopard,* Zeuzera pyrina, *is an example of such. The large white moths speckled with blue-black used to turn up frequently in Birmingham although the larvae bore into the wood of fruit trees. As a small boy I often wondered at the strange life cycles of many insects and the highly specific food requirements of many of their immature stages.*

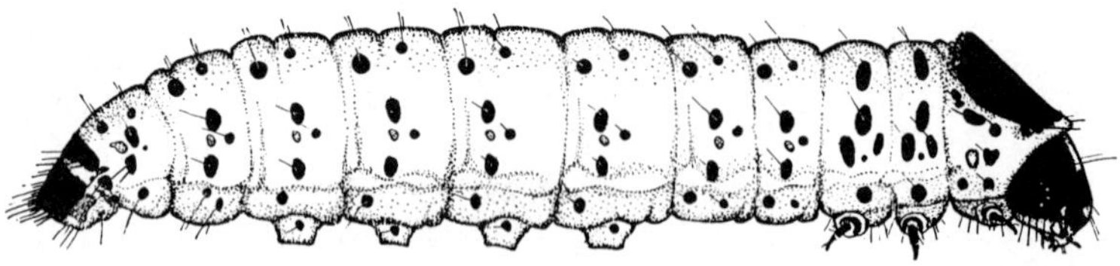

his days were numbered when the news got round, and suddenly, as if by magic, a host of reports came to hand of poultry (including turkeys) mysteriously slaughtered by a savage rogue badger. Then poor old Brock, who chose semi-domesticity in which to end his days, fell in unequal combat to the gun.

Many 'writers about plants' have called attention to the beauty of the brilliant yellow flowers of yellow mimulus. This plant is indigenous to North America but has been found wild in several parts of Surrey for many years, and how delightfully it enlivens those 'moist, shady places' where it is found. Its Bletchingley haunt is just below Tilgates. How well I remember meeting it! It was a Sunday evening and my wife and I were returning from a short walk with our two small daughters. We were feeling somewhat 'down', with a touch of loneliness maybe, when we espied the mimulus clustering round a wet clay spot where some seepage ran out from the side of a small copse. All sadness and loneliness were forgotten. We whooped, and my wife had forcibly to restrain the children from wading straight into the bog. Thus was demonstrated the power of beautiful plants to change man's thought and buoy him up.

It is often a source of wonder the unlikely places where plants often grow. I only know yellow balsam to occur in a holly hedge at Bletchingley, and no other place. Its scientific name is *Impatiens noli-tangere*, and its common name touch-me-not. The girls loved to touch the tips of the fruits, causing them to dehisce with a mild explosion, never failing to make one jump and, of course, causing the seeds to be thrown several yards from the parent plant.

Guess what! Just as I had written the last sentence my eye alighted on a seed pod of the camel's foot tree (*Bauhinia*) which I had brought back from the Congo. I picked it up and wondered if it would dehisce, so I gave it a sharp tap on the desk and my! did it dehisce! The sound was like a rifle-shot, followed almost simultaneously by the sound of the seeds striking the walls of my study. Thank goodness I wasn't in the line of fire.

The man orchis grows in great profusion in some parts of Surrey within twenty miles of St. Paul's, but it is rather local. Each year we made a pilgrimage to see it. The quaint man-like flowers never failed to thrill us. It grows within a few feet of a public footpath, but the soft yellowish-green colouration of its inflorescence is its salvation. Its charm is form, not colour.

We also liked to search for the large Roman snail (*Helix pomatia*), the edible species said to have been introduced to Britain by the Romans.

Even in our own garden, the stone boundary wall which was only a few feet from the actual tarmac of the road, we experienced many pleasures of natural history. The white admiral butterfly used to lay its eggs on the snowberry bushes *Symphonicarpus*, and it was not until I looked it up in the flora of Bentham and Hooker that I learnt that this bush is placed in the same family, *Caprifoliaceae*, as the normal foodstuff of this butterfly, the honeysuckle (*Lonicera*). The curious caper spurge (*Euphorbia lathyrus*) grew as a weed and has followed us around in our transplanted clumps of garden perennials ever since. We once had a visit from some hawfinches. These retiring birds spent a day amongst our apple trees.

It was a pleasant walk to the North Downs where we could look for many chalk-loving plants and their associated insect life. We would sit high up looking down on to the heavily wooded country and we could forget the thundering traffic along the A25, which mounted in volume as each year passed.

Within a year of working at Stockwell our little factory was almost bursting at the seams, and as it was impossible to contemplate any sort of building growth there, Miss Eades began to seek out larger premises. These were eventually found in the village of Fetcham near Leatherhead in Surrey. In fact, the building was the village dance hall, the functions of which had been transferred to the converted implement shed and barn of the nearby Home Farm. Moving our machinery, tanks and stocks was a colossal task, but at last it was finished, although at a cost of much physical exertion. This now meant a journey for me across Surrey of sixteen miles from Bletchingley to Fetcham, which was made for some years in a succession of various motor-cars. I cannot explain why it was that Home Farm remained unlived in for such a long time before I became aware of the fact, for I had admired it for some years. All the surrounding farmland had been 'developed' and the farm now stood with a long drive in front of it and a small piece of rough orchard at the back. There was a large hole in the roof, however, and water poured right through the middle of the house every time it rained. In spite of this, dry rot was absent. Conditions had obviously been much too wet for the fungus *Merulius lacrymans*, but there were some patches

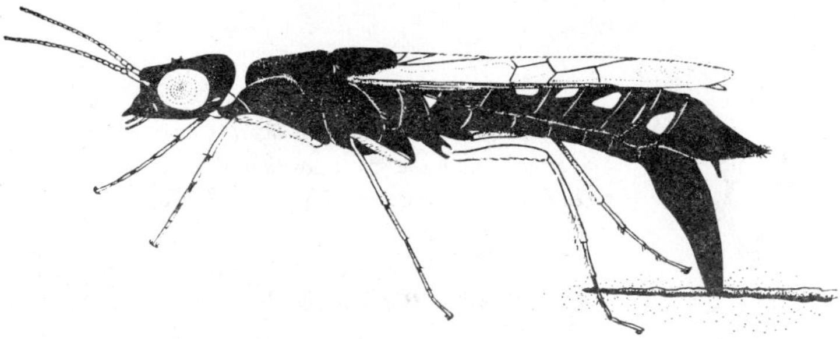

All along Dowles the alders cluster thickly, extending into tall dark thickets where the winding brook has cut oxbows. The silver birch from the mixed oak coppice on the steep banks has also migrated to the dark brookside. In this location both trees grow with tall spindly trunks. But such an etiolation saps their vigour and they fall victim to several species of wood-rotting fungi. Then on a hot summer's day the female alder wood wasp (Xiphydria camelus), *shining black with white-spotted abdomen, finds the freshly fallen trunk and, curling the tip of her abdomen downwards and inserting the strong blade-like ovipositor guards in a longitudinal crack in the bark, she cuts a channel with the sharp rasps of her egg tube down to the cambium layer and through the tiny orifice of the egg tube the eggs are extruded. At this stage the eggs may be parasitized by a hymenopteron,* Aulacus striatus. *The females of the latter wait about around the egg-laying wood wasp and when she has withdrawn her egg-laying tube, the* Aulacus *quickly inserts her own down the channel and then passes an egg into any wood wasp egg that, by this means, she can reach.*

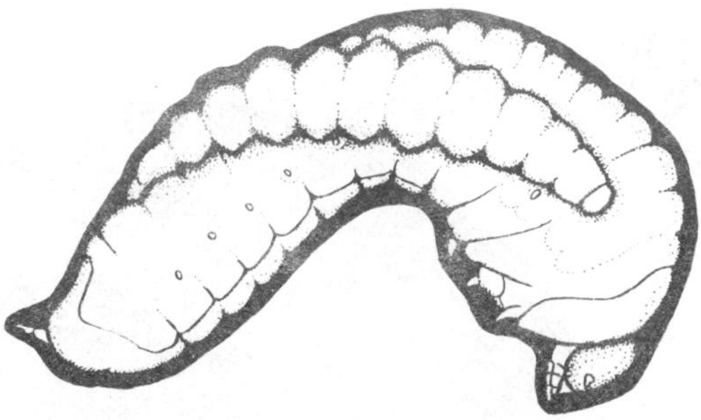

It could almost be assumed that the larva of the alder wood wasp, Xiphydria camelus, *burrowing in the dying trunks of alder and birch along the banks of Dowles would be safe from further parasitization after* Aulacus striatus *had planted its brood in some of the wasp eggs. This is not so, as at least three other hymenopterous species are specially adapted to prey upon it.* Rhyssella curvipes, *an ichnuemon, by some sense can detect the nearly full grown wood wasp larva inside the wood and with its thin needle-like egg-laying tube can bore a hole direct to the larva and pass an egg down on to it. It hatches almost straight away and after hooking itself on to its host it sucks its juices so rapidly that within sixteen days the unfortunate wood wasp larvae is a bit of dried skin. The* curvipes *larva then lives in suspended animation for ten months before it assumes the pupal stage. Everything is so carefully timed that it does not bore out of the wood until the following year's wood wasp larvae are nearly full grown for the pattern to be repeated.*

All trace of the old charcoal hearths have now gone. It is only because I know the exact position of a number of them that I can examine their position. The succession of mosses and flowering plants that grew on these black circles was studied in Wyre Forest by those two great botanical ecologists E. J. Salisbury and A. G. Tansley. But now the climax has nearly been reached and a few young oak saplings are found amongst the dense forest of silver birch scrub, that in turn took the place of the rosebay, willowherb and the sunrose. During the last twenty years the Forestry Commission have covered one or two tracks with black cinders thus making a similar ecological habitat to the charcoal circles, and it is at the edge of the cinder tracks that the saw-wort grows. This plant is like a small version of the greater knapweed and no doubt its specific name Serratula tinctoria *refers to the fact that a good yellow dye can be made from it.*

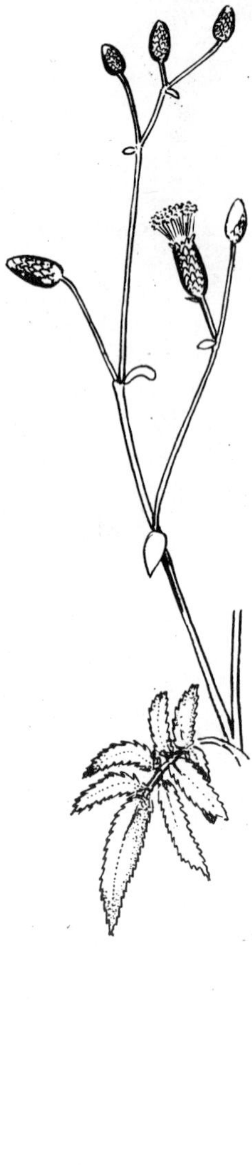

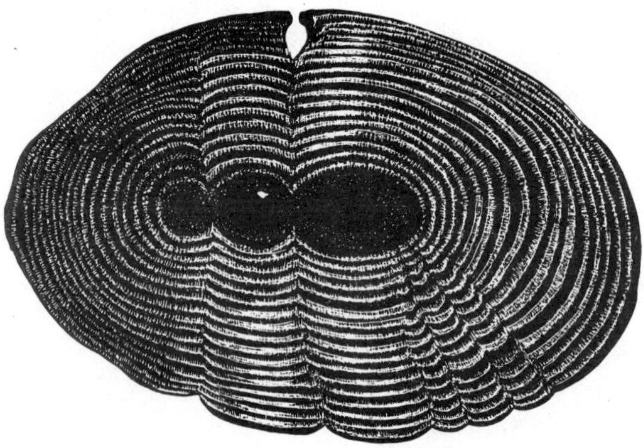

If we cut the sporophore of Daldinia concentrica *we see the concentric pattern of the alternating growth zones. Although not yet proven it seems reasonably certain that alder and birch must be attacked by this species of fungus before the alder wood wasp,* Xyphidria camelus, *larvae can feed on its wood. If there is any parallel between the biology of the alder wood wasp and the wood wasps of the conifers* (Sirex *and* Urocerus), *then the larvae of the alder wood wasp will possess pouch-like organs actually storing special mycelial tissue of* Daldinia.

The trees of the forest fight for their place in the light, and, ultimately, they fall to the axe and the saw, but here and there trees, mis-shapen or wounded by storm or fire, fall to the wood-rotting fungi. Most of the latter, whose spore-bearing organs are large enough to strike our gaze, belong to the BASIDIOMYCETES *group whose spores are produced in fours on little stalks, but the fungus* Daldinia concentrica *belongs to the* ASCOMYCETES. *In this group the spores are produced in vase-shaped cells in sixes or eights.* (Vase = ascus.) Daldinia concentrica *strikes the ash most commonly and the silver birch and alder more rarely. We show the black bulbous sporophores on a slender trunk of ash.*

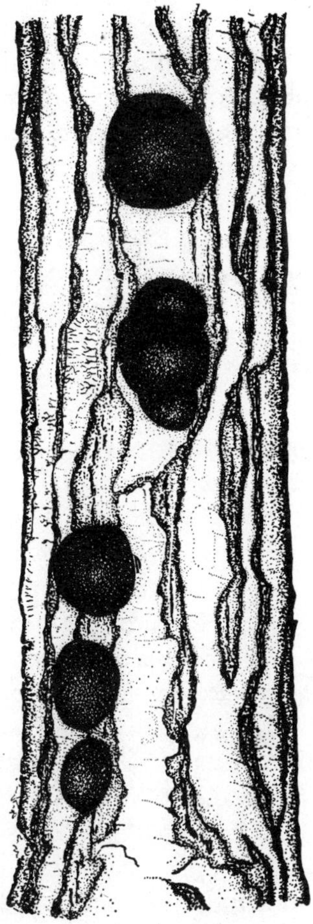

of wet rot due to the fungus *Coniophora cerebella*. On the other hand, woodworm attack by *Anobium punctatum* was widespread; in fact in some timber, especially in the roof, the infestation was severe. One survey report described Home Farm as unfit for human habitation on account of woodworm.

After I had made a detailed survey I offered the owners £1,200 for the property and to my great surprise this was accepted. Thus I was able to get a major conversion of a very old property, which was built in 1667, out of my system. Emma had successfully put me off the conversion of Legge's shop in Bletchingley, but she realized, I suppose, that this was something I had to do some time in my life. Nine years have now passed since I preserved the timber, removed the worst of the Victorian partitions and grates and decorated throughout. It was a job of considerable magnitude to bring it into some semblance of modern living requirements. Within the four walls there did not exist a pipe or a wire of any description, so that we literally had to start from the beginning in electrical and plumbing matters.

The fact that so many of the timbers supported woodworm infestations did not worry me very much, although in the first summer the little brown beetles were flying around all over the place. There was a lot of timber to cover and even now I have not got around to treating all of it, but the incidence of the beetle got less and less until scarcely one is found today. The steel-blue beetle *Korynetes caeruleus* that preys on woodworm was also present at Home Farm, but strangely it is resistant to insecticidal fluids and completes its life cycle, despite the fact that the piece of wood from which it emerges has been liberally dosed with woodworm-killing fluids. There was so much wood in the construction that I removed a large quantity of rotted beams and ceiling joists and replaced none.

Appropriately it was at Home Farm that I wrote my book *The Insect Factor in Wood Decay*, and, with the exception of those done at Wyre Forest, all the drawings, too. I would return home from Felcourt, my company's headquarters on the Surrey/Sussex borders near East Grinstead, and after a tête-à-tête dinner with Emma, our two daughters now being at university, I would prepare my desk for an evening's drawing. The specimen, usually a beetle, was taken from an insect-cabinet drawer and placed under the special binocular microscope mounted on a long arm which is pivoted on to a heavy base, and thus has the great advantage of giving ample room for drawing materials as well as the specimen under the optical system. I would then draw and write descriptions of the specimens until I was too tired to continue, whilst Emma sat a few feet away and we would chat from time to time until we went to bed. I was most happy when spending an evening thus.

10

The Forest Old and New

WHAT was the forest like in former times? What sort of people dwelt around its fringes? These are questions which come into the mind when scrambling along the little streamlet valleys that empty into the Dowles.

In the latter quarter of the sixth century the west Saxons driving northwards up the Severn Valley after their decisive victory of Dynham in 577, which gave them Gloucester, Bath and Cirencester, were stopped in the northern half of what is now called Worcestershire, but was then called Wireceastrescire or some slight variant. The rolling hills, thickly forested, came right to the Sabrina Bank, giving it a gorge-like aspect as can be seen to this day, just south of Arley where Seckley wood on the west bank and Eymore wood on the east give a vivid impression of what an impenetrable barrier the forest was to settlement in those times. This was the Forest of Wyre, in those times probably stretching to the east as far as Clent and the Lickey Hills, and to the west as far as the Clee Hills in neighbouring Shropshire. This naturally difficult countryside preserved a tribe, the 'Hwicce', dwelling precariously in Gloucestershire, Worcestershire and west Warwickshire, from their enemies in the north and from settlement by the Saxons in the south. Historians are able to tell us this on account of place-names research, because there are very few names of heathen burial settlement places. Even the earliest burial places have a Christian origin.

The Forest of Wyre, the county and the county town have a common root. It was the *foresta de Wira* in 1177, but this is thought to have its origin in an earlier name, Wigra. This also has significance as a river name and scholars have conjectured whether this was an earlier name for the present Dowles Brook which now cuts the modern forest in two. In 1350 the town was called *Wircestre* and in 1459 even *Wyrettur*. There is no doubt, however, that the forest spread over a great part of the north of the county and was a formidable area.

Much of the Worcestershire part of the forest is in the parish of Rock, the name having an interesting derivation. At first sight, and in fact in most modern inhabitants' minds, it would have some significance with regard to the substance rock. But that is not the case; far from it being mineral, it is in fact vegetable. In the thirteenth century there are several references to this parish as Ak, which is the Anglo-Saxon word for oak. Indeed, there are a number of allusions to a well-known meeting place under an oak; at this date there is also a spelling—Oke. In the fourteenth century it began to have a prefixed R. Then the spelling Rokke and Roke followed, with the meaning 'at the oak'. Although for some centuries now, the confused spelling of Rock has been universally employed, etymologists consider that the modern form should be Roke.

It was about the year 602 that St. Augustine met representatives of the ancient British Church 'under an oak'. It was the custom in those times for meetings to take place out of doors, and trees well known as landmarks, and sufficient to give a little shelter from the weather, were always chosen. Scholars have thought that the St. Augustine meeting was either at the 'Mitre' oak near Hartlebury or at 'the oak' at Rock. Both trees have now gone but it is known that the 'Mitre' oak was a descendant of the 'Rock' oak, so that it seems fairly certain that St. Augustine had his British meeting in the Wyre Forest of those times, and, indeed, only a mile from the present forest's edge. Today, as one comes along the road from Clows Top to Pensax, Rock Church stands on a hill as though on an island of fields, but then the ground drops away and one of the best views of the forest is seen, the thick rounded green cut up not only by Dowles but by all the little streamlets which drain into it.

At first it might seem strange that there exists little record of what was happening in the forest during medieval times. We know that it was used for the chase and that many large trees were felled. Great areas of the forest were nibbled away and where suitable brought under the plough. This process, of course, goes on to this day, but during the last few years has slowed down somewhat on account of the reversal; some areas that had gone from rough grazing to gorse and heather having been replanted with spruce and larch. But in the late seventeenth century the town of Bewdley on the Severn, at the south-east corner of the present area, became a bustling town. It was a port of some consequence and its great stone-slab quays are there now. Old prints show large-sailed trows loading and squared lumber awaiting shipment. There must have been more water in the river then, perhaps as much as two feet, as such craft could not now get afloat except in times of flood.

The forest then provided much of the raw material for Bewdley industry. When the forest became coppiced, allowing the oak to spring again from the old cut stumps, it was able to provide large quantities of wood for fuel, charcoal and oak bark for tanning. Until just before the Second World War bark could still be seen being stripped from the long oak poles, and the charcoal-burners, too, went about the forest building their cottage-sized fires. Along almost all the forest tracks the old black rings where a charcoal mound has smouldered can still be detected.

Side by side with the coppicing existed many rural crafts; the splitting of oak staves and weaving the lathes which were produced into large baskets and trugs was quite important. One old craftsman died only recently but I remember him steaming the oak staves and splitting them; it seemed so easy. I do not know when the last coracle was used on the Severn at Bewdley but I remember seeing a man using one between Bewdley and the Blackstone Rock. He was raising his long wicker salmon traps. I was ten years old.

The tanyard of William Beacall at Bewdley still tanned leather in 1932 and a leather works still occupies the site. It should be mentioned that the forest was often called Bewdley Forest and I was present as a youth when Stanley Baldwin, the Prime Minister and Member for Bewdley, was presented with the freedom of the borough and with a large collection of examples of rural crafts of the district. It seemed to me, however, that all these things, the woven baskets and the like, were the remnants of an age fast slipping away.

The end of the eighteenth century and the beginning of the nineteenth saw a change in the forest. This was the building of a large number of small cottage-type farmhouses around the forest. The great farming boom was taking place. Corn-growing on an extensive scale in all those parts of the country where corn could possibly be grown pushed cattle-grazing into the forest lands.

These small farmhouses are a characteristic feature of the edge of Wyre around the whole of its perimeter and, in addition, they may be found, even if now only a pile of stones, deep in the heart of the forest. I almost wrote that only their 'foundations' can be found, but this would have been a figure of speech, because usually they did not have any.

The life of the mole is one of feverish activity. It has to be, because a few hours' starvation, when it cannot find worms, causes death. An interesting point concerning its worm catching, however, has emerged during the last few years. This concerns the worm stores which the mole makes when worms are plentiful and it has eaten enough. Each worm is bitten at the front and in such a way that the worm is partially paralysed and cannot escape. As many as a thousand worms are sometimes found together in such a store. Where do these worms come from? The most likely explanation is that the worms, which are usually of only one species, are immobilized by a sudden frost.

The weasel is perhaps the commonest mammal seen to cross the highway, at least that is the writer's impression gained during the last twenty years of extensive motoring. It does so, always exactly transversely, whether the motor car is almost on top of it or not. Its gait, when crossing the road, is an exceedingly fast run, its little legs moving so fast that they can scarcely be seen. Its normal progression over rough grassland is a series of short bounds, yet it is never seen to bound on to the road. It would appear that it suddenly finds itself on the roadside through travelling along mouse and vole tunnels, which is its normal hunting procedure, and then it feels it must go on and not turn back.

The slabs of bunter sandstone were laid straight on to the ground or on to a few larger rocks. Over the windows and doors red bricks were used as though the stone was not to be trusted when keyed into an arc. Bricks were also used for the top course to support the common rafters of the roof. They are, then, of many colours, for the sandstone has weathered to a variety of mustard yellows and the red bricks stand out like eyebrows over the windows. The ground floor is of brick laid straight on to the ground, and the floor and ceiling joists of the upper rooms are of roughly squared oak poles of about five inches across and laid at about two feet between centres. This means that when woodworm has taken off the outer sapwood you may not dance in the bedroom. Partition walls were also of oak poles, but considering how easy they must have been to obtain the very minimum were used. They can easily be seen from their curvature to be coppiced poles, growing from an old butt. There was not much accommodation, as fully one-third of it was given over to stands for two cows with a hayloft over.

The small farmsteads were usually of only a few acres in extent and many of their little fields were on steep hillsides. They were well planted with fruit trees and the cherry and damson did very well, although only the cherry now has any significance as an economic crop.

What sort of people are the forest people of Wyre? This again is a difficult question to answer because as country people living in isolation they have been largely unsung. Unfortunately, they do not come well out of what writing there has been.

For instance, Noake, writing in 1877, said:

'In former years the Far Forest was a nursery for thieves and ruffians and used to furnish more victims for the gallows than all the country beside. The late Rev. J. Cawood had much to do with civilizing the scattered populations, and bringing them within the happier influences of religious ministrations. In medieval time, when Bewdley was a sanctuary town, and the neighbouring forest was a hotbed for highwaymen and murderers, this must indeed have been a neighbourhood most undesirable for residence, notwithstanding its great natural beauties.'

It was the minister mentioned above about whom it is still told that he would give a pig or a calf to a forest couple if they would consent to be married in church.

But George Jorden, a famous local botanist, always seemed able to move about the forest without hindrance, and following his early years, when the great number of Victorian naturalists of various sorts explored and collected in the forest, botanists, geologists, entomologists and students of other creatures, one is never conscious of anything but quiet seclusion and isolation from their writings.

In some ways I envy George Jorden who commenced to tramp and clamber his way through the forest before the end of the eighteenth century. He was a remarkable man in the way that he triumphed over his beginnings. He was born near Farlow in 1783 on the Titterstone Clee Hill. His father was a labourer, but his mother had a reputation for her knowledge of herbs and their curative properties, and this obviously had some influence on young George. When he was quite young he went to Bewdley as an errand boy, but during this time he taught himself to read and write. Living in Bewdley at that time was a cultured and eminent physician, James Fryer, and he invited Jorden to work for him, dispensing medicines and generally helping in the practice. Jorden did this for fifty years, but he was able to spend a great deal of time in the forest, collecting plants, and he became a well-known figure in the botanical world. He rediscovered the famous and unique Sorb tree, the Whitty Pear as it is called today. He invariably rose before dawn and went rambling in the forest, returning with many specimens for his herbarium which he preserved with great care. He called this his *Flora Bellus Locus* and it is now in the Worcester Museum. His records of plants in Wyre Forest made significant contributions to Leighton's *Flora of Shropshire* and Lees' *Botany of Worcestershire*.

Jorden died in 1871 at the age of eighty-seven. He was certainly forest refreshed. Walking and scrambling through the forest in his day, of course, was a real physical effort, as there were no main tracks like the modern Forestry Commission firebreaks. What paths there were had been made by forest workers to get the timber out; a season's work followed by half a century's disuse. Moreover, such timber tracks, as all modern forest-walkers and all who have been lost in the forest will know, strike out in great arcs, often coming back to the original point and always changing course so that one's sense of direction is seldom of help after a mile.

One of the great changes that have taken place in the forest during the last hundred years has been the disappearance of the 'Great Bog'.

This I have searched for, but no vestiges remain, nor do any present forest folk have any knowledge of it. The 'Great Bog' of Wyre Forest was said to be the highest point of Bewdley Forest and George Roberts in 1860 tells us to turn aside from Dowles Brook where Park Brook falls into it and a woodcutters' track leads you to it. These seem rather strange directions to the highest point of Bewdley Forest (that is, the near part of Wyre rather than the Far Forest), for this is not a great distance from the road from Bewdley to Cleobury Mortimer and it would have been an easier walk from this road along which were enough inns to provide a landmark.

But it was at the Great Bog that the July air was scented with the fragrant orchid (*Gyamnadenia conopsea*) and where the very rare summer lady's tresses (*Spiranthes aestivalis*) was found. At its edges Jorden, Lees, Matthews and other nineteenth-century enquiring and exploring botanists, found flowery treasures such as wintergreen (*Pyrola minor*), birds' nest orchid (*Listera nidus-avis*), marsh helleborine (*Epipactis palustris*), lily of the valley (*Convallaria majalis*), the mountain melic grass (*Melica nutans*) and purple molinia grass (*Molinia caerulea*).

It is improbable that the present-day botanist will find the night-scented summer lady's tresses, as it has not been found in Wyre Forest since 1854 and flowers in July and August, now only in one or two bogs in the New Forest. But the diligent searcher should find the beautiful sword-bladed helleborine (*Cephelanthera longifolia*) as a few plants are to be found where they were first observed a century ago, but they are associated with beech, not oak. It is also still a great thrill to see lily of the valley flowering in the forest as an English native plant.

It is worth a note here that the true fragrant orchid, G. *conopsea*, is found on chalk downs and limestone pastures and not in marshy ground. The marsh fragrant orchid, G. *densiflora*, however, is found in such situations, together with *Epipactis palustris*. The fragrant orchid of Wyre Forest is, therefore, almost certainly the marsh fragrant orchid of the later writers. The green-winged orchis (*Orchis morio*) was quite common in the riverside pastures around Dowles Church where it occurred with the early purple (O. *mascula*) and the cowslip. When I was a boy this was an unforgettable sight. It can now only be a memory as they have been picked to extinction. Jorden was the botanist who found the fragrant lady's tresses (*Spiranthes spirialis*), the narrow-leaved white helleborine (*Cephelanthera longifolia*), the broad-leaved helleborine (*Helleborine latifolia*), marsh helleborine (*Helleborine longifolia*, also

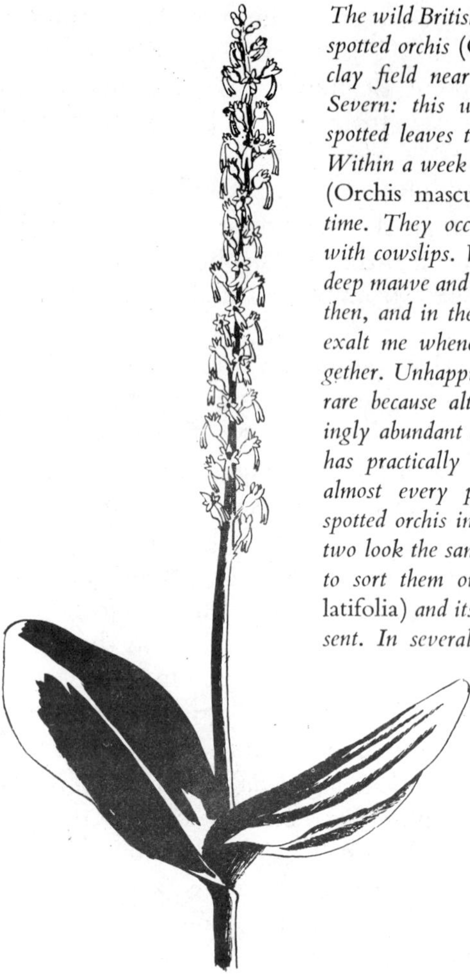

The wild British orchids fascinated me from boyhood. The spotted orchis (Orchis maculata) *was my first find in a clay field near where Dowles empties itself into the Severn: this was when I was nine years old. The spotted leaves thrilled me with their exotic appearance. Within a week or so I found a field full of early purples* (Orchis mascula), *the bulls' pissles of Shakespeare's time. They occupied a Severn bank meadow together with cowslips. What a delight the contrast of colour, the deep mauve and the lime green and pure yellow was to me then, and in the intervening years it has never failed to exalt me whenever I have seen these two flowers together. Unhappily, as the years go by, this becomes more rare because although the cowslip has become increasingly abundant around the forest the early purple orchis has practically disappeared. Today, within the forest, almost every patch of marshy ground supports the spotted orchis in great quantity and in large variety; no two look the same, but I have never got around to trying to sort them out to see if the marsh orchis* (Orchis latifolia) *and its hybrids with the spotted orchis was present. In several places along the railway embankment through the forest another orchid, twayblade,* Listera ovata, *was common when I was a small boy, and today is found in abundance at the Newalls and in many of the small rough fields on the banks of Dowles. This is the subject of the illustration.*

called *Epipactis palustris*), summer lady's tresses (*Spiranthes aestivalis*), and the frog orchid (*Habenaria viridis*) in the forest, whilst Matthews found the large butterfly orchid (*Habenaria virescens*), and *Orchid inarnata*. Perry found the fragrant orchid (*Gymnadenia conopsea*) and others found the birds' nest orchid (*Neottia nidus-avis*), common twayblade (*Listera ovata*), the marsh orchis (*Orchis latifolia*) and the lesser butterfly orchis (*Orchis bifolia*).

Jorden's collection of plants has been in the museum at Worcester since 1909, and they are still wrapped in the brown-paper packets labelled by him in the old classification. There are 814 specimens, very dusty, but fairly well preserved for all that. Not all are marked with the locality of collection, but others are evocative of Jorden's interests and times such as, 'taken from the walls of Ludlow Castle', 'Titterstone Clee, near the devil's armchair' and 'Dowles, near the vitriol works'. The latter has now disappeared without a trace, but was situated near the outflow of the brook into the river Severn. When I went through his collection in 1963 I could well have been the first to unfasten the packets since 1909 and look through the specimens, comparing the species he found in the forest in those days with those that are there today. Two of the orchids in his collections, birds' nest, *Neottia nidus-avis* and the fragrant orchid, *Gymnadenia conopsea*, I have not observed within the forest, although I am familiar with them both in other parts of England.

When writing of George Jorden of Wyre Forest and the plant life that he discovered it is natural to follow with an account of the Whitty Pear, a unique tree discovered or rather rediscovered by him.

The minute pygmy shrew, Sorex minutus, *hunts with feverish activity by day and by night; the spells of about one and a half hours of hunting are followed by sleep of a similar period. Every twenty-four hours it consumes something like three-quarters of its own weight of insects, worms and other small creatures, fresh flesh of recently dead mammals with some vegetable matter such as the seeds of forest trees. The lactating female eats twice as much. If this food is not eaten regularly it dies of starvation within a few hours. The adult weighs from 3 to 6 gms, that is from 9 to 5 of them weigh an ounce. It somehow seems strange that its total length of life does not take in two winters. The pygmy shrew is a tremendous climber and is known to crawl up the rough bark of trees in its diligent search for victuals, and it is perhaps this habit that gets it sometimes to the upper floors of a house. This little animal is common around the forest and the specimen illustrated was found in a bedroom at the Newalls.*

In early spring there flowers in some of the forest's bordering fields a plant which I have never identified with satisfaction. Rather like the ladysmock, its pale lavender flowers stand about a foot high, but they are double and there are neither stamens nor pistil. Reproduction is obviously vegatative, small plants springing from the tuberous root stock. Is this coralroot, Cardamine bulbifera? *If so it bears little resemblance to the illustration in Bentham and Hooker.*

When the tawny owl has digested the night's catch of voles and other small rodents, it coughs up the indigestible remains as a large pellet. The fur of its prey covers the outside and serves to lubricate the sharp bones and skulls. They are almost as large as the animals it gulps down, two and a half inches long and nearly an inch in width being quite usual dimensions. The cry of the tawny owl is the last sound heard in the forest before sleep comes.

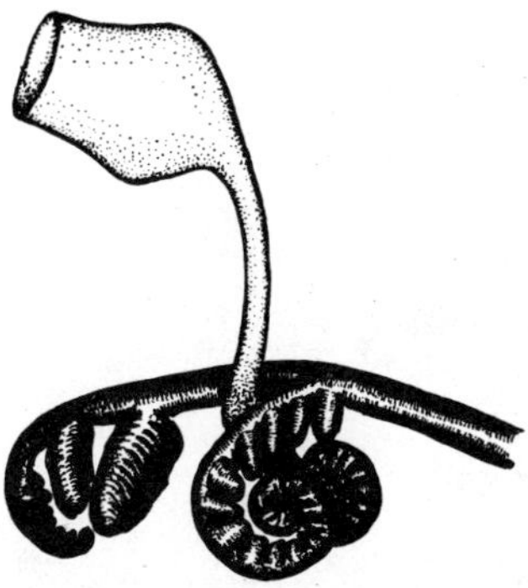

The Agroeca *spider spins an egg cocoon rather in the shape and size of a bilberry and attaches the long stalk to nodules of loose soil. The illustration shows one which had been attached fortuitously to the tightly folded coil of a bracken pinnule under the soil surface and when at the end of May the frond shot upwards it took the* Agroeca *egg case with it to breast height, where rain washed the soil particles away and revealed this pearl-coloured object in an incongruous position.*

John R. Burton, Rector of Dowles, writing in 1883 in his history of Bewdley in Worcestershire, said that 'a sorb tree' (*Pyrus domestica*), apparently the only wild tree of the species in Britain, grew 'till lately' in the forest. It was burnt down by some miscreant in 1862, but perhaps the incendiarist was not wholly to blame, perhaps he thought he was the victim of some witchcraft in which a branch of this notorious tree took some part. How long has this tree or its ancestors stood in this glade? It is mentioned in the *Philosophical Transaction* for 1678 by an Aldermann Pitts of Worcester and it was considered an old tree at that time. There are illustrations of the tree in Nash's *History of Worcestershire* and in Lees' *Botany of Worcestershire*. The ruins of a hermitage were traced out by George Jorden near where the tree stood and a gold coin of the Emperor Tiberius was found in the forest in 1783 not far away from it. The sitting figure with a spear holds a branch of olive.

But a fine Whitty Pear still stands to this day. Whatever powers for good or evil it possesses, sufficient people wish to go and seek it out to make worth while the erection of a signpost. From Ford Lane Crossing, on the railway line, walk towards Bewdley on the line a few hundred yards to Park House, a forester's bungalow where a lately cut and broad forest track runs up from Dowles Brook to the Bewdley–Cleobury road. Twenty minutes' walk up this road (about a mile) to the south along its wide sweeps through the forest brings you to a turning to the left where stands the sign. In fact, where the track turns off in a sweep to the right, the Whitty Pear track goes straight on. Some way on another track cuts across and a forester's hut has been built there, and a water hole dug, but if you keep on the original direction

The Whitty Pear, then called Sorbus domestica *as illustrated in* Collection for the History of Worcestershire *by T. R. Nash, 1781. This is the same tree as described by Edmund Pitt in the Philosophical Transactions for 1678 (Vol. 12), entitled 'Account of the* Sorbus pyriformis'.

Leaves and fruit of the Whitty Pear as illustrated by Nash, 1781. Pitt in 1678 said of the fruit 'In September so rough as to be ready to strangle one. But being then gathered, and kept till October they eat as well as any medlar'.

I am grateful to the Society of Antiquaries for permission to reproduce this and the preceding figure.

the way soon leads to the Whitty Pear in a broad glade between conifer plantations. The O.S. grid reference is SO/748748.

The first impression is of a stout-boled tree, much larger than a mountain ash which its leaves so closely resemble. In shape it is a rounded pyramid, but the lower branches are dead, perhaps through having been snatched at too frequently by those who know its history of magic.

In modern floras such as that of Clapham, Tutin and Warburg (1958) a small paragraph is given over to our tree, here known as *Sorbus domestica*, the Service Tree, where it states that is is larger than the rowan, with spreading branches and with a rough scaling bark. The fruits are green or brown, apple- or pear-shaped, about an inch in diameter. It also states that 'a single tree formerly grew (probably introduced) in the Wyre Forest, but it is apparently not now known as a wild tree in this country'. Its wider distribution covers the south of Europe, North Africa and Asia Minor.

How did the Whitty Pear reach the middle of a great forest? How could a seed have been implanted there in the middle of the sixteenth century, reckoning the tree that Aldermann Pitts described as a hundred years old? We can be sure that the forest was a much wilder place than it is now, wild animals abounding and the chase jealously guarded.

But there the Whitty Pear stands today with a white post in front on which is a bronze plaque bearing the following inscription:

This
Sorb Tree
a direct descendant
of the original tree
was presented by
Robert Woodward Esq.
of Arley Castle
and planted by
Mrs. R. Woodward
on 30th March 1913

John Humphreys Esq.,
M.A., M.D.S., L.D.S., L.F.S.A.,
F.I.C., F.G.S.

President
F. T. Spackman Esq., F.G.S.,
Now Secretary

The wood of the Service Tree has a reddish tint, is fine-grained and can be given a very high polish. Loudon says that the wood of the Service is the hardest and heaviest of all trees indigenous to Europe, and that the tree takes two centuries to reach its full stature of fifty to sixty feet, and 'lives to so great an age that some specimens of it are believed to be upwards of 1,000 years old'.

Its two names, the Service and Sorb, have interesting derivations, both having their origin in the drink made from its fruits. Service is thought to come from *Cerevisia*, a fermented drink, and Sorb from the Latin *sorbeo*—drink down.

The derivation of the name of the very closely related rowan tree is fascinating. It shows a connection between old Norse and Sanskrit languages. In both cases the word is *runa*, a charm in Norse and a magician in Sanskrit. Incidentally, one of its common names is Fowler's Service. In Evelyn's *Sylva* occurs the passage, also quoted by Edward Step, 'Ale and beer brewed with these berries, being ripe, is an incom-

The Whitty Pear of Wyre Forest as illustrated in Lees' Botany of Worcestershire, 1867, *as it appeared many years before but obviously in decline. I am grateful to the Linnaean Society for permission to reproduce this and the figure below.*

The Whitty Pear of Wyre Forest as it was in 1856. A Whitty Pear stands today in the position it formerly occupied. One of the few trees of England which is signposted.

parable drink.' Oh, for an old recipe! What fun to try it with the fruits of the true Service which are there for the picking in October.

What of the Wyre Forest today? What is its present extent? The answer to this question depends on whether areas of woodlands continuous with the forest 'proper' are included or not. I think perhaps that what the local forester would call the forest is in the shape of a diamond, lying within the two and a half inch to one mile ordnance map 32/77. From this map it is difficult to distinguish wooded land, as it is not coloured, and the little tree symbols are printed in light grey. Why the symbols could not be printed in green I have no idea. If the one-inch ordnance map Kidderminster Sheet 71 is consulted the larger diamond shape made by the Bewdley to Cleobury Mortimer road to the south and to the north, the road from Bewdley to Kinlet and thence to Cleobury is easily made out. Measured on the map this makes the long east-to-west axis about six and a half miles, and the greatest north to south (the diamond is somewhat lopsided) three and a quarter miles. But there is a long forest walk for those with staying power. This is to start at Arley Wood near Stourport on the Severn and to commence by walking northwards to Ribbesford and then turn west to Long Bank on the main road. The forest is entered on the north side of the road at Beaucastle and the general northwards direction is continued through Lords Yard Coppice and Town Coppice to Lodge Hill Farm where the railway and Dowles are crossed. Traversing Chamberline Wood the Bewdley–Button Oak road is crossed entering Hawks Batch and Seckley Woods. This brings us to the Severn Gorge and to continue walking through unbroken forest it would be necessary to swim the Severn because the Elan Aqueduct bridge also crossing the river at this point does not cater for the public. It means, therefore, a walk into Arley and cross the ferry; an enjoyable journey. You then get into Eymore Wood but continue north-east to Shatterford village and passing over the Kidderminster–Bridgnorth road into the last mile and a half of woodland through Birch Wood, Coldridge Wood and finish in a wood of the same name from which we started—Arley Wood. On the map this measures ten and a half miles, but the convolutions of the paths are sure to make this a fifteen-mile walk.

There is another section of the forest, rather cut off from the main woodland, by the open country of Far Forest village and the unwooded area at the back of Furnace Mill. This woodland measures two and three-quarter miles down from the north through the Coachroad Coppice and Breakneck Bank. Between the railway and the main road is Bell Coppice, whilst to the south of the main road is Cleobury Coppice, its western extremity being the deer park of Mawley Hall. This is probably the appropriate place to warn my readers against a hazard of forest-walking. After several miles of walking through oak wood accompanied only by one's private thoughts and then to step on to a main road is unnerving. One should sit for a minute on the roadside before attempting to dodge the stream of fast cars.

There are two aspects of human activity in the forest that are worthy of a few more words. Firstly, there is the railway. This is a single-line track from Bewdley to Cleobury Mortimer on the Tenbury Wells line. The river Severn is crossed exactly where Dowles empties into the river and then the line follows from east to west just south of Dowles Brook. It is possible to walk the length of the Dowles path without realizing that a railway exists. The line was opened in 1861 and was closed to passenger traffic exactly 100 years later to the day. It is destined to close altogether in a couple of years and no one knows what will then happen to it. Actually, the cutting and bank have provided habitats for unusual plants, such as the lesser toadflax (*Linaria minor*). This little four-inch annual with a rather dusty appearance was first found in 1883 by Lees at Wyre Forest railway station and it was thought to be new and a description was published. It was subsequently found not to be so, but to be a native of Southern Europe, having been introduced here. Eighty years have now passed and it is still found at Wyre Forest station; the gritty, well-drained paths at the side of the ballast obviously provide it with ideal growing conditions.

The second human activity concerns coal. Bridgnorth in the north and the Abberly Hills in the south marks an extensive coal field, the Forest of Wyre coal field. In area it is about the same as the Forest of Dean coal field, but in mineral wealth is much inferior. Even so, two small pits, Bayton and Arley, produce coal today. It is recorded that in 1911 there were four small coal pits working.

When I first camped in the forest near the Horse and Jockey Inn a coal pit was pointed out to me which was just a hole in the ground, like a well, and an ordinary domestic bucket was lowered down on a rope some thirty feet. Recent exposures made during the construction of firebreaks in the forest have revealed several very thin black seams of coal only five or ten feet from the surface. After a period of prolonged rain, such as in the early spring, the wet surface clay, which is of a light mustard or fuscous colour, is so tenacious that one could easily lose one's shoes if they are loose fitting. But the tracks in the forest have often been worn down by the iron-tyred wheels of timber wagons down to the shales and bunter sandstone that lies beneath the poor cladding of soil.

The fossils of the carboniferous age lying around the wafer-thin seams of coal were well known to the Victorian geologists. They found about twenty species of ferns in the shale which splits up so easily and oilily, the more important being *Sphenopteris affinis* and *S. bifida*, *Neuropteris lashii*, *N. gigantea* and *N. accuminata*. A number of

Pecopteris species are there too, including *P. adiantoides*, *P. lonchitica*, *P. serlii* and *P. oreopteridis*. A horsetail is also present, *Asterophyllites foliosa*, and a club moss, also a rather strange reticulate-veined fern given in books as *Woodwardites robertsi* with a question mark behind it. There is a crumbling bank on the Salop side of Dowles which has been well picked over by university students from Birmingham these last few years. Unfortunately, when these plant fossils are removed from the position which they have retained for 200,000,000 years they crack up into oily fragments. Fossils of animals seem very rare; I have never found any there.

Obviously there have been many changes in the bird-life of the Forest. The grouse went many years ago but the long-eared owl and the woodcock remain. After an absence of many years the buzzards have returned.

It is now many years since the late Professor Steele Elliot persuaded the Pied Flycatcher to nest in Wyre Forest near Dowles Manor by providing suitable but artificial nesting sites. In the year 1963, two pairs nested in an old overgrown orchard near Weavers on the brook, their nearest neighbours being in the Forest of Dean. As we stood immobile against an old damson tree in the hedge, a cock, in his black and white, delighted us by fluttering down from a moss-covered crab apple and, hovering like a humming bird, picked a fly from a buttercup and then returned to his perch. The beauty of this little action was most exhilarating and I find myself recalling it to mind and reliving the breathless thirty seconds.

As I now walk through the forest I am sure that I must cut a somewhat forbidding figure with a little green beret pulled over one side of my closely cropped head and my great heavy figure draped with an old tired and torn tweed coat. My rucksack is of a venerable vintage having served me for nearly forty years, and although the canvas is now beginning to rot it is nevertheless still capable of carrying a half-hundredweight tree-stump back to the Newalls. The mustard-coloured dust or mud of the bunter clay sticks about me as I pull myself up a bank to gaze at and admire an old rotting coppiced stump of oak. The durable heartwood often twisting and writhing assumes patterns of great beauty which affect me greatly. After absorbing this sight I test the weight and calculate what reserves of strength remain in me at the particular stage of my walk and whether I am equal to the task of carrying it back to Newalls so that I can see it more often.

It is seldom that one meets people in the forest. At Easter a few Scouts camp down by Dowles, but to be confronted by strangers in the woods and coppices is a rare experience. Yet much biological research and collecting has been accomplished in recent years and it seems to make the forest even more remote to read in current scientific journals of a piece of research carried out in Wyre that one knew nothing about. This has even occurred when the author was a man I knew. For a few years I sat on the Council of the Institute of Biology with W. B. Yapp of the Department of Zoology of Birmingham University, yet I did not know until recently that he had mapped the distribution of the willow warbler and the wood warblers in Wyre. He found that the latter bird predominated in the natural oak copses whilst the former was found in the regenerating silver birch.

The foresters are fewer. When I was a boy the woodcutters worked in teams of three and lodged in cottages near the scene of the cutting and they spent their evenings grinding their axes on a stone that they took around with them. Today one hears the whine of a petrol-driven hand saw and I have seen a man cutting by this means sizable trees singlehanded, with his little open estate car parked on a nearby track.

Of the three species which go to make up what we call English oak, that of Wyre Forest is known as the durmast oak with the specific name *petraea*, although the old Worcestershire botanists referred to it as *intermedia.* It is characterized by the leaves having long stalks and being downy underneath, whilst the acorns have only short stalks. The other two species forming the oak 'complex' are *sessiliflora* in which the acorn stalk is very short but the leaf stalk is from half to one inch in length, and *pedunculata* in which the leaf stalk is absent or very short whilst the acorn stalk is very long. Furthermore, in this latter species the leaf has a distinctive shape, being broad at the base, this giving it a heart-shaped appearance. The species *sessiliflora* predominates in the north and west, perhaps the greatest present-day stand being the Forest of Dean. *Pedunculata* belongs to the south and is the oak of Sussex, of the forests of Worth, Tilgate and Ashdown.

Next to the oak and the silver birch the beech is increasing in importance. There are some very large beech trees in Wyre, for instance in Oxbind coppice, but in large areas, especially around New Parks, beech is creeping into the oak. Of course, beech is well able to do this on account of endurance to shade, but when once it gets the upper hand its leaf mosaic is so perfect that nothing will grow underneath it, so well does it shut out the light. Some of the beech has been planted, but in other places it has obviously seeded down. Caesar's statement that there were no beech trees in Britain was accepted for many centuries, but it is now known for certainty that it thrived in many counties of England when the Romans came.

Many yews remain in the forest and many times have I sheltered from the rain amongst their fluted and twisting roots. Some of them that are near my frequented paths are like protecting friends and there is one by the island amongst the roots of which we hid our kettle and frying pan from one week's end to the next.

The tremulous aspen and the blowing gean (in the forest blowing is pronounced to rhyme with ploughing) are in the forest too, but how can I describe in the course of a few words all that goes to form the perpetual forest?

H. L. Edlin, the forester, recently wrote: 'These are the trees we love, the trees that we shall remember, and the forest that they form is a living forest because it calls upon and inspires the dreams of men and women for its full understanding.' I suppose that a certain feeling of satisfaction is given to me by Wyre because of its ancient character. The dominant tree is the oak, the durmast oak, *Quercus petraea*, and the second most common tree is the silver birch, *Betula pendula*. During the Ice Age, that lasted until 10,000 years ago, Britain had no trees but there still existed the land bridge with Continental Europe, so that when the ice receded the land was invaded by trees again. It is known that the birch was the pioneer and this was by virtue of its remarkable seed. One pound of it contains half a million seeds and the trees produce vast quantities year after year. Such fine seed is blown about by the wind long distances. But, although birch is the first tree invader, it cannot hold its position except in rather exceptional situations, and the oak can become established because the birch not only does not cast a dense shade but as a tree is rather short-lived. At eighty years the birch polypore has sent its mycelium deep into its heart and its white brackets appear high upon the trunk. Then comes the time of the oak. What a remarkable thing it is to contemplate that the oldest vestige of oak in our islands is some pollen of about 8,000 years ago, following the retreat of the ice, and yet today oak is the commonest tree in our country, representing one-third of all our trees of the woods and hedges. Oak is about seventy-five years old before it bears its fruit,

The narrow-leaved helleborine (Cephalanthera longifolia=ensifolia) *is still to be found in the forest. Thirty-four plants were found on the edge of new timber tracks in New Parks on the Worcestershire side during the first days of June 1963. The fourteen flowers of the largest specimen were pure lily white with the lip of the labellum bright yellow and its base tinged with flesh tint. This species is much rarer than large white helleborine,* C. latifolia, *from which it differs in having only one or two visible bracts at the base of the twisted carpels, as well as the leaves being long and narrow.* Longifolia *also is fertilized by insects and its flowers are generally opened widely instead of being self-pollinating and the flowers hardly opening as in the large white.* Longifolia, *as Jocelyn Brooks has observed, has all the appearance of a tropical orchid that has strayed into an English wood. George Jorden found this species which was then called* Cephalanthera ensifolia *over a hundred years ago in the same New Parks area and he also found the broad-leaved helleborine* Cephalanthera latifolia *which I have not found there. There is some confusion in Bentham and Hooker's* British Flora *in the account of this plant due to the name* ensifolia *being cited as a synonym of* latifolia.

M

the acorns, which can only drop and bounce from the tree. Its average rate of distribution by this means alone has been calculated as only one mile every 2,000 years. Obviously we must look for another means of distribution which has given us in a few thousand years our national tree. Perhaps it was the wood pigeon, filling its craw with acorns and flying some miles before being struck down by a falcon.

It seems certain that Wyre Forest has existed in its present form of oak and silver birch since the ice retreated. Although the planting of conifers makes some difference to the general aspect of the forest, European larch, Norway spruce, Scots pine, Corsican pine and Douglas fir have all been planted extensively during the last quarter-century and many of the stands show considerable promise, although they are dwarfed by the giant redwoods at Goodmoor. Along the brook the plantations of individual species are small so that in any one scene one gets a glance of two or more tones of green. With an old yew hanging on to an escarpment and with honeysuckle and the alder buckthorn fighting for light at the plantation edge the scene is altogether a beautiful one.

What is it that compels me to return to the forest as often as my energy permits? What makes me drive each Friday evening 140 miles across the width of England? It is difficult to analyse one's feelings about this. I know that when I am not in the forest, when it is possible for me to be there, I am moody and irritable and life seems very noisy.

I long for the dappled sunlight lighting up the cow wheat on the mossy oak banks. I long to see whether the delicate mauve stars of the autumn crocus have been pushed through the summer's fading stalks by their long white perianth tubes. I want to see whether the black strangalia beetles are yet congregating on the yellow-green cyathia of the wood spurge and I want to ponder on its biology which no man yet knows. When the hot July sun comes I imagine I see the glossy black flash of the white-barred clearwing as it tastes aerial flight after imprisonment in the alder stems, and I can see it in the dimly lit thicket with hover flight searching the alder bark.

I would want to see whether the larvae of the terrestrial caddis, *Enoicyla*, that for nearly a hundred years have been seen in Britain only by my family and I, are still sheltering under last year's oak leaves that the winter's storms have blown into the grassy tunnels of the long-tailed field mouse.

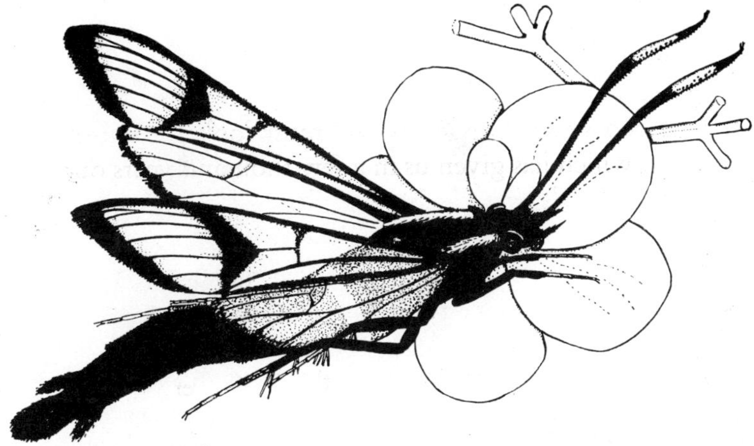

It was when searching for the black strangalia longhorn beetles on the flowers of sun spurge, where they congregate, that I saw the first white-barred clearwing moth. It was a large freshly emerged specimen and it circled around me first and as it did so I hadn't the faintest idea what it was—it just looked large and black. Then it settled only four or five feet away from me and only a few inches from the ground on the branch of a very small seedling spruce tree. The moment it settled I could see that it was the very rare white-barred clearwing (Aegeria spheciformis). *The wings are diaphanous clear but with marginal black scaling at the tips and in prominent moon-shaped patches. Around the black silky body was a girdle of large yellowish-white scales with two diagonal creamy braces over the shoulders. But strangely the most prominent part of the insect were the yellowish areas near the tips of the antennae. Observing it I was able to bring my face within twenty inches of it when without any preliminaries it suddenly flew off.*

*The winter of 1961–2 had been long and cold and spring so delayed that the spring flowers were still in full bloom when the summer flowers were unfolding. At the edge of the young spruce trees the bluebells made June 17*th *seem like two months earlier. But there had been ten days' hot sun—the last few days very hot and this was the day of emergence of* spheciformis.

The following day I watched a female spheciformis *searching the alders at the edge of Dowles for suitable egg-laying sites. She travelled up young alder streams hardly flying but hovering like a humming bird. In the shadow cast by the alder leaves at midday the wings were invisible, they moved so rapidly, but the yellow antennal patches and the girdle were quite distinct and served as a pattern so that the movements of the insect could be followed with ease in the gloom. Two or three years in the life cycle of this insect are spent as a larva boring down young alder shoots, until it emerges in the morning of a hot fine day, to sun itself for a time and to make some proving flights in the outside world. To be present in the forest at the emergence time of this rare and beautiful creature was a most stimulating experience such that even though later that day I drove one hundred and forty-four miles across England to my home in Surrey I was forest refreshed.*

References

I am sure it must be obvious to the reader that during the writing of this book, which has occupied me off and on for many years, I must have consulted many books, papers and standard textbooks concerning the illustrations and the plants and animals that I have written about. It did not seem appropriate to include all these in this list of references. However, I thought it might help or interest the reader if I included a short list of works which I have consulted most recently or which, because of their special nature, I keep near my hand.

Amphlett, J., and Rea, C., *The Botany of Worcestershire*, Cornish Bos. (Birmingham, 1909)

Brooke, J., *The Wild Orchids of Britain*, Bodley Head (London, 1950)

Browne, E. O., and Burton, J. R., *Short Stories of the Worthies of Worcestershire*, E. G. Humphreys (Worcester, 1916)

Chamberlain, E. L., *A Glossary of West Worcestershire Words*, English Dialect Society, Truber (London, 1882)

Crowcroft, P., *The Life of the Shrew*, Max Reinhardt (London, 1957)

Edlin, H. L., *The Living Forest*, Terames and Hudson (London, 1958)

Greene, S. W., and Clark, M. C., 'The Bryophytes of the Wyre Forest', *Proceedings of the Birmingham Natural History and Philosophical Society*, **20** : 3–22 (1962)

Laird, N. D., *A Topographical and Historical Description of the County of Worcestershire*, Sherwood, Neely and Jones (London). (The beauties of England and Wales.)

Lees, E., *Botany of Worcestershire*, Worcestershire Naturalists Club (Worcester, 1867)

Leighton, W. A., *Flora of Shropshire*

Mansergh, E. L., and W. L., 'The Works for the supply of water to the city of Birmingham from Mid-Wales.' Minutes of *Proceedings of the Institution of Civil Engineers* (1912)

Mawer, A., and Stanton, F. M., *The Place-names of Worcestershire*, University Press (Cambridge, 1927)

Nash, T. R., *Collections for the history of Worcestershire*, 2 vols. (London, Oxford and Worcester, 1781)

Noake, J., *Worcestershire Relics*, Longmans (London, 1877)

Reiss, F. A., *The History of the Parish of Rock Worcestershire*, H. Grant (London, N.D.)

Roberts, G. E., *The Rocks of Worcestershire, their mineral characters and fossil contents*, Joseph Masters (London, 1860)

Rolt, L. T. C., *Worcestershire*, Robert Hale (London, 1949)

Salisbury, E. J., 'The Vegetation of the Forest of Wyre: A preliminary account', *Journal of Ecology*, **13** : 314–21 (1925)

Sankey, J., 'How to Begin the Study of Harvest Spiders', *Countryside*, **17** : 1–8 (1956)

Symonds, W. S., *Records of the Rocks; or, Notes on the Geology, Natural History and Antiquities of North and South Wales, Devon and Cornwall*, John Murray (London, 1872)

Wells, H. G., Huxley, J., and Wells, G. P., *The Science of Life*, Cassell (London, 1931)

Yapp, W. B., *Fauna from Birmingham and its regional setting*. A scientific survey (1950)

Yapp, W. B., 'Ecological Speciation in the genus *Phylloscopus*'. *Proceedings of the Xth International Ornithological Congress*, 173–5 (Uppsala, 1951)

Yapp, W. B., *Birds and Wood*, Oxford University Press, (London, 1962)

Index

A

B

C

V

W

X

Y

Z